T0336133

VOLUME ONE HUNDRED AND THIRTY THREE

ADVANCES IN
COMPUTERS

Internet of Things: Architectures for
Enhanced Living Environments

VOLUME ONE HUNDRED AND THIRTY THREE

ADVANCES IN
COMPUTERS

Internet of Things: Architectures for
Enhanced Living Environments

Edited by

GONÇALO MARQUES
Polytechnic Institute of Coimbra,
Technology and Management School of Oliveira do
Hospital, Oliveira do Hospital, Portugal

ACADEMIC PRESS

An imprint of Elsevier

ELSEVIER

Academic Press is an imprint of Elsevier
125 London Wall, London, EC2Y 5AS, United Kingdom
525 B Street, Suite 1650, San Diego, CA 92101, United States
50 Hampshire Street, 5th Floor, Cambridge, MA 02139, United States

First edition 2024

Notices
Knowledge and best practice in this field are constantly changing. As new research and experience broaden our understanding, changes in research methods, professional practices, or medical treatment may become necessary.

Practitioners and researchers must always rely on their own experience and knowledge in evaluating and using any information, methods, compounds, or experiments described herein. In using such information or methods they should be mindful of their own safety and the safety of others, including parties for whom they have a professional responsibility.

To the fullest extent of the law, neither the Publisher nor the authors, contributors, or editors, assume any liability for any injury and/or damage to persons or property as a matter of products liability, negligence or otherwise, or from any use or operation of any methods, products, instructions, or ideas contained in the material herein.

ISBN: 978-0-323-91089-7
ISSN: 0065-2458

For information on all Academic Press publications
visit our website at https://www.elsevier.com/books-and-journals

Publisher: Zoe Kruze
Editorial Project Manager: Palash Sharma
Production Project Manager: James Selvam
Cover Designer: Greg Harris

Typeset by STRAIVE, India

Contents

Contributors

Qammer H. Abbasi
James Watt School of Engineering, University of Glasgow, Glasgow, United Kingdom

Leocundo Aguilar
Universidad Autónoma de Baja California, Baja California, Mexico

Adnan Nadeem Al Hassan
James Watt School of Engineering, University of Glasgow, Glasgow, United Kingdom

Fehaid Alqahtani
James Watt School of Engineering, University of Glasgow, Glasgow, United Kingdom

Shuja Ansari
James Watt School of Engineering, University of Glasgow, Glasgow, United Kingdom

Kamran Arshad
James Watt School of Engineering, University of Glasgow, Glasgow, United Kingdom

Khaled Assaleh
James Watt School of Engineering, University of Glasgow, Glasgow, United Kingdom

Silvio César Cazella
Federal University of Health Sciences of Porto Alegre – Department of Exact Sciences and Applied Social – Information Technology and Healthcare Management, Rio Grande do Sul, Brazil

Sergio Cervera-Torres
Multimodal Interaction Lab. Leibniz-Institut für Wissensmedien (IWM), Tübengen, Germany

Alejandro Dominguez-Rodriguez
Psychology, Health, and Technology, University of Twente, Enschede, The Netherlands; Health Sciences Area, Universidad Internacional de Valencia, Valencia, Spain

Maitreyee Dutta
National Institute of Technical Teacher's Training and Research, Chandigarh, India

Muhammad Farooq
James Watt School of Engineering, University of Glasgow, Glasgow, United Kingdom

Madushika Gamage
Department of Computer Science and Engineering, University of Moratuwa, Moratuwa, Sri Lanka

Hanyang Hu
School of Architecture, Tsinghua University, Beijing, China

Miloslav Hub
University of Pardubice, Faculty of Economics and Administration, Institute of System Engineering and Informatics, Pardubice, Czech Republic

Muhammad Ali Imran
James Watt School of Engineering, University of Glasgow, Glasgow, United Kingdom

Muhammad Zakir Khan
James Watt School of Engineering, University of Glasgow, Glasgow, United Kingdom

Hana Kopackova
University of Pardubice, Faculty of Economics and Administration, Institute of System
Engineering and Informatics, Pardubice, Czech Republic

Bee Theng Lau
Faculty of Engineering, Computing and Science, Swinburne University of Technology
Sarawak, Sarawak, Malaysia

Jie Liu
School of Digital Media and Design Arts, Beijing University of Posts and
Telecommunications, Beijing, China

Dan Luo
School of Architecture, University of Queensland, St Lucia, QLD, Australia

Gonçalo Marques
Polytechnic Institute of Coimbra, Technology and Management School of Oliveira do
Hospital, Oliveira do Hospital, Portugal

Dulani Meedeniya
Department of Computer Science and Engineering, University of Moratuwa, Moratuwa,
Sri Lanka

Analúcia Schiaffino Morales
Federal University of Santa Catarina – Sciences, Technologies and Health Education Center
– Computer Department and Graduate Program in Energy and Sustainability, Santa Catarina,
Brazil

Violeta Ocegueda-Miramontes
Universidad Autónoma de Baja California, Baja California, Mexico

Fabrício de Oliveira Ourique
Federal University of Santa Catarina – Sciences, Technologies and Health Education Center
– Computer Department and Graduate Program in Energy and Sustainability, Santa Catarina,
Brazil

Indika Perera
Department of Computer Science and Engineering, University of Moratuwa, Moratuwa,
Sri Lanka

Adnan Qayyum
James Watt School of Engineering, University of Glasgow, Glasgow, United Kingdom

Sanka Rasnayaka
School of Computing, National University of Singapore, Singapore, Singapore

Antonio Rodríguez-Díaz
Universidad Autónoma de Baja California, Baja California, Mexico

Jagriti Saini
AMIE, IEEE Member, Mandi, Himachal Pradesh, India

Mauricio A. Sanchez
Universidad Autónoma de Baja California, Baja California, Mexico

Rosario Sánchez-García
Universidad Autónoma de Baja California, Baja California, Mexico

Ione Jayce Ceola Schneider
Federal University of Santa Catarina – Sciences, Technologies and Health Education Center – Health Science Department, Santa Catarina, Brazil

Ahmad Taha
James Watt School of Engineering, University of Glasgow, Glasgow, United Kingdom

Muhammad Usman
James Watt School of Engineering, University of Glasgow, Glasgow, United Kingdom

Sandareka Wickramanayake
Department of Computer Science and Engineering, University of Moratuwa, Moratuwa, Sri Lanka

Weiguo Xu
Institute of Future Human Habitat, Shenzhen International Graduate School, Tsinghua University, Beijing, China

Pan Zheng
Department of Accounting and Information Systems, University of Canterbury, Christchurch, New Zealand

Preface

This book is the 133rd in the *"Advances in Computers"* book series, which has been published since 1960. This volume focuses on the emergent advances in computing technologies based on the Internet of Things (IoT) for Enhanced Living Environments (ELE) and aims to provide an in-depth review of the latest research findings and technological developments in this research field.

In Chapter 1, entitled "Explainable artificial intelligence for enhanced living environments: A study on user perspective," Wickramanayake et al. discuss how ELE applications utilize the expressive power of Artificial Intelligence (AI) to provide enhanced performance. Despite the impressive performance, state-of-the-art AI techniques come at the cost of explainability; users cannot understand the rationale of the decisions made by such systems. This chapter addresses user perception and requirements of eXplainable AI (XAI) and attitude toward adopting explainable ELE systems. The conducted user study revealed that most perceive XAI as essential and expect explanations to be easy to understand and interactive. Moreover, this study presents a novel approach to generating multimodal explanations consisting of linguistic and visual explanations to rationalize the decisions made by Human Activity Recognition systems. The conducted validation survey highlights that users' trust grows with explainability, leading to higher adoption of ELE systems.

In Chapter 2, entitled "Human behavioral anomaly pattern mining within an IoT environment: An exploratory study," Sánchez-García et al. state that standard psychometric tools, such as questionnaires, are the gold standard for assessing clinical disorders but face significant drawbacks regarding subjective bias. On the other hand, the development of IoT networks has been part of the technological advances that are characteristic of Industry 4.0 due to the large amount of information provided by networked sensors regarding the environment and the interaction of individuals in it, allowing the detection of behavioral patterns exercised. This chapter presents an innovative approach to assessing mental health by taking behaviors as the primary input. The general IoT sensor network was designed in-house to enable high control over the flow of information related to target conducts. The results show that detecting anomalies within human behavioral patterns is possible. Therefore, an ELE system could be potentially applied in contexts such as suicide prevention or discovering other undiagnosed mental disorders that individuals may present throughout their lives.

Chapter 3, entitled "Indoor localization technologies for activity-assisted living: Opportunities, challenges, and future directions" by Khan et al., provides a comprehensive analysis of indoor localization, including challenges, solutions, and technological advancements. It thoroughly examines different conventional and machine learning-based approaches and techniques and their pros and cons, along with evaluation metrics such as accuracy, cost, and energy consumption. It offers solutions for cost-effectiveness, energy optimization, and overcoming environmental obstacles. Furthermore, it outlines major industry stakeholders and the evolving market landscape, highlighting the sector's growth and emphasizing the necessity for further research, integrating diverse methodologies, and developing various indoor tracking technologies. This chapter's insights are relevant for researchers, practitioners, and enthusiasts engaged in this field, bridging academia and industry and offering significant value across multiple domains.

Chapter 4, entitled "Smart indoor air quality monitoring for enhanced living environments and ambient assisted living" by Saini et al., focuses on the importance of indoor air quality, which has been a major concern for public health and well-being. People who spend 80%–90% of their routine time indoors are likely to experience serious complications due to repeated exposure to degraded air quality. Patients with disability and existing health concerns usually interact more with the indoor environment. Therefore, the risk of rising concentration of air pollutants in the building premises in urban and rural areas poses a considerable threat to people with assisted living. This chapter focuses on the efforts made by past researchers, potential gaps in the literature, future scopes, and technical recommendations to deal with the problem domain. This chapter also introduces the latest technologies, methods, and approaches to support ELE and ambient assisted living.

Chapter 5, entitled "Usability evaluation for the IoT use in Enhanced Living Environments" by Kopackova and Hub, presents the status of usability engineering methods in the development of ELE using IoT technologies. This chapter shows that the usability evaluation of IoT technologies in the ELE context is evolving, but it is still at its beginning. The presented literature review reveals that most studies tested the usability of ELE technology by the end users. It also shows that none of the evaluated studies explicitly justified the choice of the usability evaluation method. Additionally, none of the studies under analysis compared the performance and suitability of various usability engineering methods by simultaneously applying different methods to the same problem and comparing the outcomes.

Chapter 6, entitled "Internet of Things to enhanced living and care environments for elderly: Applications and challenges" by Morales et al., focuses on the importance of considering the social and cognitive aspects of ELE for aging individuals. This chapter provides a roadmap for this area, investigating it through a rapid review of 58 papers on IoT-based systems developed in the last decade. There are still few studies tested with older people as volunteers, leaving several gaps in how these technologies can gain social acceptance among aging people. The concept of monitoring daily activities was discussed in 18 papers. Although several sensors were deployed to gather information about the daily activities of seniors, transmit it to the cloud platform, and synchronize it with caregivers or other identified individuals, privacy and security considerations were not taken into account. Several challenges in caring for senior adults were identified, indicating that questions about senior adaptability with technological resources, privacy, and security issues deserve further attention in future works.

Chapter 7, entitled "Internet of things and data science methods for enhanced data processing" by Zheng and Lau, focuses on the challenge of handling the massive data produced by IoT systems. IoT is an expanding domain involving the interconnection of physical objects embedded with sensors, software, and connectivity. With a growing number of IoT devices, there is a proportional stream in the data they generate. This abundance of data holds promise for extracting valuable insights and facilitating improved decision-making across diverse fields. Nonetheless, the volume, velocity, and variety of IoT data pose substantial challenges for conventional processing and analysis methods. This is where the significance of data science methodologies emerges. This study explores established frameworks and comprehensive methodologies extensively used in the field of data science and analytics. This work also introduces the Data Science Life Cycle for IoT Applications (DSLC-IoT), which serves as a fundamental guideline to tackle the challenges stemming from data intensity and dependence in the context of IoT. Subsequently, this work analyzes various data science approaches employed to enhance data processing within the IoT context.

Finally, Chapter 8, entitled "Internet of things challenges and future scope for enhanced living environments" by Liu et al., explores the challenges and future potential of integrating the IoT to enhance living environments. It addresses technical, social, and economic challenges, emphasizing the pivotal role of smart architecture and cities in shaping the future of IoT. A central focus is supporting the construction of an intelligent IoT ecosystem that goes beyond individual components, encompassing smart products, homes, architecture, and cities. The chapter promotes the integration of IoT within this ecosystem, covering technical aspects such as data collection, integration, advanced analytics, and real-time monitoring, as well as social and economic factors such as improving user experience, efficiency, and security. In summary, this chapter provides valuable insights into how IoT can be harnessed for the improvement of living spaces.

In recent years, several new application areas for ELE based on IoT and other computing-based technologies have emerged. Therefore, this book aims to synthesize recent developments, present case studies, and discuss new methods. Numerous methods for ELE are available in the literature. Furthermore, the application of these methods is of utmost importance for enhanced public health. Therefore, it is imperative to merge emergent computer advances with IoT to develop novel solutions and solve complex challenges that contribute to overall public health and well-being.

I hope the chapters in this work serve as a relevant source of information for the development of future ELE systems based on IoT, capturing the readers' interest. Furthermore, I want to express my sincere gratitude to the contributors and all the people involved in the editing process.

GONÇALO MARQUES

CHAPTER ONE

Explainable artificial intelligence for enhanced living environments: A study on user perspective

Sandareka Wickramanayake[a] (iD), Sanka Rasnayaka[b] (iD), Madushika Gamage[a] (iD), Dulani Meedeniya[a] (iD), and Indika Perera[a] (iD)
[a]Department of Computer Science and Engineering, University of Moratuwa, Moratuwa, Sri Lanka
[b]School of Computing, National University of Singapore, Singapore, Singapore

Contents

Advances in Computers, Volume 133
ISSN 0065-2458
https://doi.org/10.1016/bs.adcom.2023.10.002

Abstract

Enhanced Living Environment (ELE) applications utilize the expressive power of Artificial Intelligence (AI) to provide enhanced performance. However, despite the impressive performance, state-of-the-art AI techniques come at the cost of explainability; users cannot understand the rationale of the decisions made by such systems. This opaque nature causes a lack of trust leading to less adoption of ELE systems, especially in mission-critical domains such as healthcare. The AI community has proposed various eXplainable AI (XAI) techniques to rationalize AI decisions. However, the user perspective of XAI is not well explored. This chapter addresses user perception and requirements of XAI and attitude toward adopting explainable ELE systems. A user study with 326 participants revealed that most perceive XAI as essential and expect explanations to be easy to understand, faithful, and interactive. The respondents prefer concept-based explanations over feature attributions. Hence, we develop a novel approach to generate multimodal explanations consisting of linguistic and visual explanations to rationalize the decisions made by Human Activity Recognition (HAR) systems. Finally, we conduct a validation survey to evaluate the impact of introducing explanations into a HAR system. The results highlight that users' trust grows with explainability leading to higher adoption of ELE systems.

1. Introduction

Enhanced Living Environment (ELE) technologies aim to create safe environments for assisted people to experience an independent and active lifestyle [1]. ELE applications utilize the expressive power of Artificial Intelligence (AI) techniques, particularly Deep Neural Networks (DNN), to provide better performance [2]. For instance, state-of-the-art Human Activity Recognition (HAR) systems for smart homes are powered by Deep Learning techniques [3,4]. While state-of-the-art AI techniques demonstrate impressive performance, they do come at the cost of explainability. Due to the back-box nature of these models, the end users do not expose to the rationale for the decisions made by the AI systems or do not realize the process being used for such decision-making. Such opaque nature causes a lack of trust, leading to less adoption of AI-powered ELE systems, especially in dependable application domains such as healthcare. For example, medical experts may use a less accurate yet understandable system over a highly accurate black-box system [5]. Further, the black-box nature obstructs troubleshooting the systems, leaving developers unaware when systems make wrong decisions. Hence, it is imperative to devise methods to explain the decisions made by AI systems.

Existing techniques to rationalize AI decisions, also known as eXplainable AI (XAI) techniques, can provide explanations in different modalities such as visualizations [6], linguistic explanations [7] and rule sets. Most of the existing XAI techniques have been predominantly algorithm-centered, with less consideration for user requirements. Since ELE applications directly interact with the public and may be deployed for dependable application domains, it is vital to understand the user requirements of XAI in ELE applications before developing new XAI techniques or adapting existing ones.

Motivated by the gaps mentioned above in the existing work, this chapter investigates the user perception and requirements of XAI and attitude toward adopting explainable ELE systems. The objective of this research is to inform the researchers of the design considerations to increase the adoption of ELE systems in the real world. A user study with 326 participants reveals that despite the demographic differences, most participants think knowing the rationale of AI decisions is essential. Moreover, the participants have indicated that explanations for AI decisions should be easy to understand, faithful, and interactive. ELE applications in mission-critical domains such as healthcare and security have been ranked higher for those requiring explainability.

Based on the findings of this user survey, we propose a new method to generate multimodal explanations for Human Activity Recognition (HAR) systems that have diverse applications in ELEs, such as medical and disability assistant applications and security. These multimodal explanations consist of linguistic explanations and visual explanations. Finally, by adapting Technology Acceptance Model (TAM), we conduct another user study to verify that users are more likely to adopt HAR systems that provide explanations in addition to the decisions. As another novel contribution, we adapt TAM to evaluate the user acceptance of an explainable DNN-based system. The analysis of the survey results has shown the importance of integrating XAI into ELE applications for the increased adoption of ELE applications.

2. Related work

2.1 AI for the enhanced living environment

Advancements in low-cost sensing technologies, wireless communication, cloud computing, and data analytics have paved the path to the employment of AI in ELE systems [8]. AI-based models have been used in almost all fields including commercial [9], social [10], environment [11], medical [12] and other human activity recognition smart application [3] domains. Assisted living systems for elderly and disabled people are another area that uses AI for decision-making in ELEs. A study focused on monitoring patients by detecting abnormal behavior has applied the YOLO CNN model as the detector [13]. They have trained the model to identify eight abnormal activities to alert the caretakers when patients have emergency medical attention. Studies have shown that wearable sensor-based activity detection can be used in the rehabilitation process for patients [14]. Vision-based activity recognition systems for smart homes are another application of AI for ELE [3].

Accordingly, the usage of AI in these ELE systems has enabled new applications and improved performances, although the underlying AI algorithms used in these systems behave as black boxes. As a result, users cannot understand the rationalization of the decisions made by these systems, which leads to obstructing the adoption of AI-based ELE systems in the real world, especially in safety-critical and dependable domains such as healthcare and security. These concerns have been addressed by integrating XIA techniques with these AI systems.

2.2 Explainable AI

Feature attribution methods and concept-based explanation methods are widely used for explaining the rationale of DNN decisions. Feature attribution methods explain a decision by indicating the influence of each input feature (e.g., an image pixel) on the model decision [15]. The feature attribution method calculates a score for each input feature indicating the influence of that feature on the predicted class using perturbation or gradient backpropagation. Depending on the data modality, calculated feature attributions are presented as a set of numerical values (e.g., for tabular data) or overlaid on the input to provide a visual explanation (e.g., saliency maps for image data).

In contrast to feature attributions, concept-based explanations rationalize AI decisions using representations easily understandable to humans, such as super-pixels or word phrases. Based on the type of explanations provided, existing concept-based explanation methods can be categorized as visual explanations [16,17] and linguistic explanations [7,18]. For example, Automatic Concept-based Explanations in Ref. [17] proposed automatically extracting image regions (concepts) important for a particular class. On the other hand, the FLEX framework proposed in Ref. [18] explains that an image is classified as a "deer" due to the concepts such as "horns" and "white dotted skin." Prototype-based explanations also have attracted attention recently, especially in the computer vision domain [19]. These methods explain image classification decisions by pointing out the prototypical parts of images from the predicted class.

2.3 Explainable AI for enhanced living environment applications

Most of the existing work in the ELE domain incorporating XAI has mainly focused on saliency maps for image data and feature attributions for tabular data. Among several studies, Le et al. in Ref. [20] proposed an explainable intruder detection system using ensemble trees. The explanations are generated using the SHapley Additive exPlanations (SHAP) [21] methodology and provide the essential features contributed to each prediction. In another study [22], explanations were given as saliency maps for driver assistance systems based on deep learning models. Another area in ELE is health status and well-being monitoring. A Gradient-weighted Class Activation Mapping (Grad-CAM) based interpretable model has been used to analyze brain tumor images in [6]. A SHAP-based explainable model for identifying subjects with autism is presented in Ref. [23], where the features that contributed to the prediction are given as explanations. In contrast to the methods mentioned above, Luca et al. in Ref. [24] propose linguistic explanations for deep learning-based activity recognition tasks in smart homes. Further, there are several medical image analysis applications with interpretable visual explanations [25] and environmental monitoring applications such as the flood monitoring system proposed in Ref. [26] with rule-based explanations. Table 1 summarizes some of the latest applications that have used XAI techniques. However, generating linguistic explanations to rationalize models' decisions based on low-level time series data such as IMU data is not well explored.

2.4 Human-centered explainable AI

Developing human-centered XAI techniques has recently attracted attention in the XAI community. For instance, a user study conducted by Kaur et al. in Ref. [27] has investigated the usefulness of the existing XAI methods to understand the processing of machine learning (ML) models. Langer et al. in Ref. [28] have identified the main stakeholders of XAI

Table 1 Summary of work incorporating XAI to ELE applications.

Application	Data type	XAI technique	Explanation type
Brain Tumor Analysis [6]	Images	Grad-CAM	Visualization
Intruder Detection [20]	Tabular	SHAP	Feature attributions
Driver Assistance [22]	Images	XRAI	Heatmap
Autism Spectrum Disorder Prediction [23]	Tabular	SHAP	Feature attributions
Activity Recognition [24]	Images	Grad-CAM	Linguistic
Flood Monitoring [26]	Images	Expert knowledge	Rule sets

and their desiderata via a literature review. They have proposed a conceptual model highlighting the relationship between XAI techniques and stakeholder desiderata. Riveiro and Thill [29], have hypothesized that factual explanations should be given when the system's output matches the user expectations and counterfactual explanations otherwise. They have conducted an empirical user study to test this hypothesis. The study has revealed that factual explanations are practical when system output matches the user expectation, and neither is useful when there is a mismatch. Another conceptual framework to develop human-centered, decision-theory-driven XAI has been proposed by Wang et al. [30], to codesign an explainable clinical diagnostic tool for intensive care phenotyping with clinicians. Even though existing literature in the XAI community has conducted experiments to demonstrate the perceived importance of XAI by domain experts and AI system developers [27,30], the perception of the users of XAI is not well explored.

3. User perceptions and requirements of explainable AI

Analyzing existing literature reveals that the perception of the users of XAI is under-explored. Fig. 1 shows the overall process view of our study consisting of the pre-survey, proposed system, post-survey and recommendations. This section explains the conducted user survey to identify the user perceptions and requirements of XAI. We hypothesize that the perception of XAI is influenced by the users' individual differences owing to the diverse audience of AI systems. Hence, we investigate the influence of demographics (gender, age, education, occupation, and career level) and AI awareness of the users on the perceived importance of XAI.

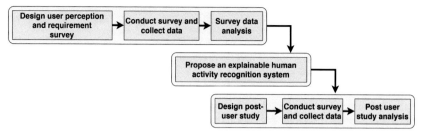

Fig. 1 Overall process of user perception study for Explainable ELE.

AI-based decision-making ELE applications and their impacts are considered critical, making XAI an essential requirement in such contexts. Initially, we explored the application types, and their effects raise the pressing necessity for explainability. We categorize existing explanation methods based on their type of explanations, such as feature attributions [15], concept-based [16], example-based, prototype-based [19] and rule-based [26]. Also, recall that feature attributions are presented as numerical values or visual explanations depending on the data modality [15]. Most of the existing studies in XAI have focused on developing explanation techniques for ML experts, developers, or specific domain experts. However, few works have recently highlighted the importance of presenting explanations in terms humans find understandable such as in natural language [31]. Hence, as a novel contribution, we explore whether the users prefer linguistic explanations over other types of explanations such as feature-attributions and prototype-based explanations.

Moreover, we investigate the characteristics that expect by the end-users from compelling explanations. There is little consensus on the notion of a compelling explanation among existing XAI work. Instead, different works have proposed explanation-generation algorithms based on various desiderata without inquiring about the end-user requirements. For example, Hendricks et al. [7] have claimed that explanations rationalizing an image classifier should be *image-relevant* and *class relevant*. In contrast, Wickramanayake et al. [18], have argued that explanations should be *intuitive, descriptive,* and *faithful.* The desired properties of explanations proposed by Ghorbani et al. [17], can be named as *meaningfulness, coherency,* and *importance.* Further, Alvarez-Melis et al. [32], have hypothesized that explanations should be *explicitness/intelligibility, faithfulness* and *stability.* Accordingly, we formulated the following five research questions and six hypotheses. The research questions

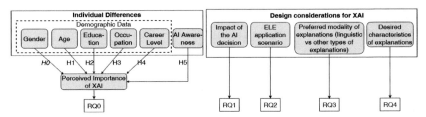

Fig. 2 Conceptual model for the user perception and requirement study.

are addressed by the individual differences of the users and by considering the identified design considerations for XAI as shown in the conceptual model in Fig. 2.

- **RQ0:** Is it important for users to know the reasons behind the decisions of AI systems?
- **RQ1:** Which impact (financial, medical, legal, etc.) caused by AI decisions gives rise to a higher importance of XAI?
- **RQ2:** Which ELE application gives rise to a higher importance of XAI?
- **RQ3:** Are linguistic explanations preferred over other types of explanations?
- **RQ4:** What are the expected characteristics of explanations?
- **H0:** Gender significantly impacts the perceived importance of XAI.
- **H1:** Age significantly impacts the perceived importance of XAI.
- **H2:** Education significantly impacts the perceived importance of XAI.
- **H3:** Occupation significantly impacts the perceived importance of XAI.
- **H4:** Carer level significantly impacts the perceived importance of XAI.
- **H5:** AI awareness significantly impacts the perceived importance of XAI.

3.1 User survey design

We conducted an online survey using Google forms employing volunteers to explore and analyze the research question and to test the hypotheses. The survey was carried out in April 2022 and consisted of three main sections. The first section collects the demographic information of the participants. The next section explores the users' awareness of having AI in four specific computational applications. This section helps to get a quantitative measure of the AI awareness of the participants. Next, the participants were given a short introduction to XAI and its usefulness using a 1-min video. The final section uses Multiple Choice Questions (MCQ), Multiple Response Questions (MRQ), the Likert scale, and open-ended questions to gather user perceptions of different dimensions of XAI.

3.1.1 AI awareness

We believe AI awareness of users significantly impacts the perception of XAI and the user requirements of XAI. Hence, we asked the participants to answer four MCQs that evaluate their awareness of the underline usage of AI algorithms in day-to-day activities.

3.1.2 Explanation type

To test the hypothesis that "*users prefer linguistic explanations over other types,*" we gathered user preferences and gave a linguistic explanation and another type of explanation for four scenarios. These scenarios were selected considering the daily activities that use ML to make decisions such as patient readmission prediction, annual income prediction, fashion recommendation, and pet recognition. In order to compare with linguistic explanations, we selected widely used explanation types namely feature attributions, saliency maps, and prototype-based explanations. Since feature attributions are more intuitive with tabular data, we considered two scenarios typically involving tabular data. In contrast, for saliency maps and prototype-based explanations, we selected two scenarios involving image data as those explanations are widely used with images.

We employed four existing methods to generate explanations under each category namely SHAP [21], Local Interpretable Model-Agnostic Explanations (LIME) [33], Grad-CAM [34] and Prototypical Part Network (ProtoPNet) [19]. The SHAP force plot for the patient readmission prediction model presented by Hilton et al. [35] was used as one of the feature attributions in our survey. We used the LIME explanation presented in Ref. [36] for a prediction for the UCI Adult Income dataset. We use a Grad-CAM visualization for the fashion recommendation model [37], as the coarse-grained saliency map. The ProtoPNet is used to get prototype-based explanations for the VOC-Part dataset [38]. We created a linguistic explanation for each explanation by describing their highlighted features. We described the top three most influencing attributes for attribution methods, concepts covered by the high severity area for Grad-CAM, and concepts in the prototypes for ProtoPNet.

3.2 Data summary

In the pre-survey, the data collection led to 326 total responses. Participants were asked to provide their age, gender, education level, field of occupation, and career level. The descriptive statistics of demographic variables are shown in Table 2. For categorical analysis, we used the mode of age (32) to split the participants into two roughly equal groups of age <32 (155 participants) and ≥32 (171 participants). The field of occupation reported as "Other" in the table includes Transportation, Production, Management, and Administrative. The Career level said as "Other" in the table consists of Student, Researcher, and Engineer.

3.3 Results analysis of the pre-survey

Table 3 states the AI awareness of the participants with the average score and the standard deviation. Recall that the quiz included four MCQs. The quiz was scored from zero to four. According to the quiz scores, the participants were categorized into two clusters. The cutoff point of 3 marks was selected by observing the resulting score distribution to divide the group into roughly equal categories. The high AI awareness category consists of people who scored three and above, and the low AI awareness category consists of people who scored below three.

To identify the participant's perceived importance of knowing the reasons behind AI decisions, we used a 5-point Likert scale. The mean score for this question is 4.32, and the standard deviation is 0.77, indicating that

Table 2 Demographic distribution of the survey participants.

Variables	Distribution
Gender	Male 52.8%, Female 46.9%, Not disclosed 0.3%
Age	Mean: 31.52 years, Standard deviation: 7.31 years
Education level	High School 14.7%, Undergrad 42.6%, Postgrad 41.1%
Occupation	Computer and Mathematical: 38.7%,
	Architecture and Engineering: 17.5%
	Unoccupied: 8.9%, Education: 8.3%, Healthcare: 5.2%
	Business & Management: 5.9%, Other: 15.5%
Career level	Unoccupied: 17.5%, Entry: 12.5%, Middle: 25.3%
	Management: 17.5%, Executive: 15%, Other: 12.2%

Table 3 AI awareness of the participants.

Awareness	Participants	Average score	Standard deviation
Overall	326	2.52	0.99
High AI Awareness (HAA)	192	3.20	0.40
Low AI Awareness (LAA)	134	1.54	0.72

Table 4 Demographic effect on the perception of the importance of XAI

	Age		Gender		Education			Occupation		Career Level	
	<32	>32	M	F	HS	UG	PG	C/M/E	Other	E	S
Mean	4.3	4.3	4.3	4.4	4.3	4.3	4.4	4.4	4.3	4.4	4.2
S.D.	0.6	0.6	0.6	0.6	0.7	0.5	0.6	0.6	0.6	0.7	0.5
P value	0.312		0.0581		0.1584			0.425		0.23	

participants consider knowing the rationale of AI decisions significant. We conducted t-tests to understand if the perceived importance varies on the participants' differences. The summary of these hypothesis analyses is given in Table 4. We used symbols M, F, HS, UG, PG, C/M/E, E, S and S.D. to denote Male, Female, High school, Undergraduate completed, Post-graduate completed, Computer Science/Mathematics/Engineering, Entry level, Senior level and Standard Deviation, respectively.

The results indicate that the perceived importance of XAI does not depend on any demographic traits at a significance level of 95%. Hence, Hypotheses 0, 1, 2, 3, and 4 are not supported. The analysis of the impact of the participants' AI awareness on the perceived importance of XAI (*H5*) demonstrates that the participants with higher AI awareness had higher perceived importance of XAI compared to the participants with lower AI awareness. The t-test indicated statistically significant support for this finding ($P = 0.0324$).

Next, we explored the impacts brought by the AI decisions, whether financial, medical, legal, social or time giving rise to the higher importance of XAI (RQ1). The participants were asked to rank these five impacts on a scale of 1–5, where 1 indicates the effect that necessitates the explanations most. The participants were allowed to give the same rank for more than one impact if they wanted to. Finally, the Relative Importance Index (RII) was calculated for the five impacts. The results shown in Table 5 indicate that the participants find that providing explanations for the models

Table 5 The RII values for the different impacts an AI system can have.

Impact of the AI decision	RII value	Rank
Medical	0.2642	1
Financial	0.2496	2
Legal	0.2229	3
Social	0.1856	4
Time	0.1842	5

Table 6 The RII values for the different ELE applications.

Impact of the AI decision	RII value	Rank
A smartwatch detects Atrial fibrillation (a common heart rhythm disorder that elevates stroke risk) in stroke survivors	0.2622	1
A surveillance system detecting a suspicious person in public places	0.2560	2
A remote monitoring system in tunnels alerts maintenance requirements	0.2286	3
An intelligent virtual assistant is helping to operate home appliances	0.1933	4
A parking slot recommendation system in the city area	0.1764	5

making medical decisions is the most important, followed by financial impact. Further, according to the participants, providing explanations for the ML models making decisions that have social or timing effects is not as crucial as medical, financial, or legal. This indicates that the context of the ML model application impacts the end user's perceived importance of XAI.

Further, we consider the perceived importance of XAI in different ELE applications. The participants were asked to rank five ELE applications on a scale of 1–5, where one indicates that the scenario requires explanations the most. We use the RII to rank the scenarios like before. The results are shown in Table 6. The results are consistent with the impacts as we observe that the scenario with the medical implications being ranked first. In contrast, scenarios with minimal impacts, such as virtual assistants and parking, are ranked lower.

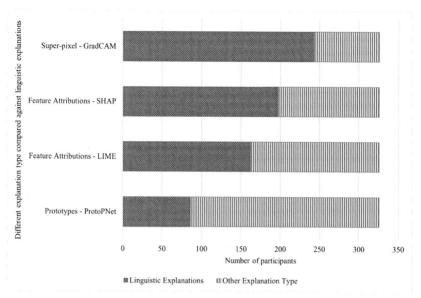

Fig. 3 Preference of linguistic vs other explanations for the different scenarios.

Next, four scenarios were used with linguistic and other explanations to understand whether users prefer linguistic explanations over other types of explanations. The participants were asked to compare the explanations provided and select what they found more helpful. Fig. 3 shows the results of this question. For Scenario 1 and Scenario 2, users have overwhelmingly preferred linguistic explanations. The choice is split in Scenario 3 and favors the visual explanation given in Scenario 4. While we see an overall trend toward preferring linguistic explanations in more scenarios, this choice depends on the scenario and explanation provided. To corroborate this association, we performed a chi-squared test between different scenarios and the linguistic vs. other choices. The chi-square statistic is 161.99, and the P-value is 0.00001, providing strong support for the association between the scenario and explanation preference.

Finally, the respondents were asked to select the expected characteristics of a good explanation. Fig. 4 shows the most important characteristics as highlighted by the respondents. The most important aspects are for the explanations to be easy-to-understand, faithful, interactive, descriptive, and consistent.

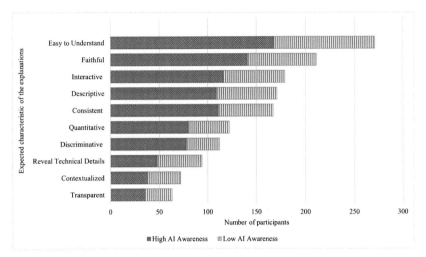

Fig. 4 Expected characteristics of explanations.

4. Multimodal explanations for human activity recognition

Human Activity Recognition (HAR) has many practical applications in ELEs, such as fall detection systems, elderly monitoring, health care monitoring, transportation, and security. Sensor-based HAR systems have recently attracted more attention due to their less intrusive and more privacy-preserving nature. State-of-the-art HAR systems employ AI algorithms to achieve higher recognition accuracy. However, these HAR systems behave as black boxes. For example, caretakers using an AI-based HAR system employed to monitor the elderly, cannot understand the reasons for the system's predictions such as the elder being in danger. Such black-box nature may hinder the user's trust in these systems. This section discusses how we can make wearable IMU sensor-based HAR systems more acceptable to users by introducing multi-modal explanations (visual and linguistic explanations) for the decisions made by the system.

4.1 Overview of human activity recognition system

Generally, human activity recognition systems consist of three main modules, as shown in the highlighted area in Fig. 5A. The data collection module collects time series data from wearable IMU sensors placed in several body locations (e.g., right wrist, left ankle, and chest). Since each body part moves

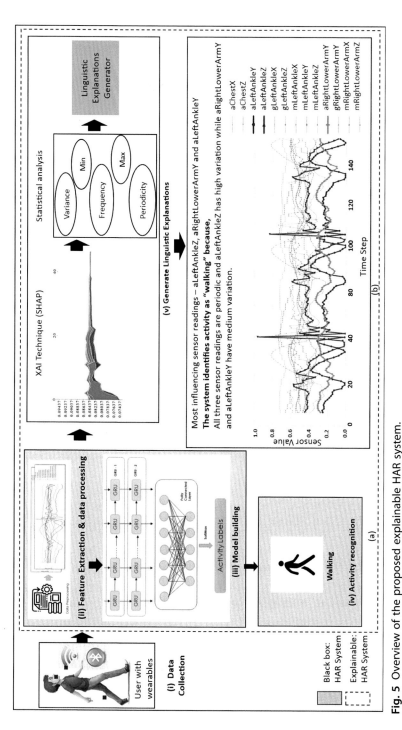

Fig. 5 Overview of the proposed explainable HAR system.

in a specific pattern for different activities, the sensor data from body-worn IMU sensors can be utilized for human activity recognition. Suppose we collect m sensor readings.

The second module, feature extraction and data pre-processing prepare the collected data for the AI algorithm utilized in the system. First, based on the mutual information statistics, we select $top-k$ features with the highest correlation with the activity classes. The goal is to remove the unnecessary features which might not be helpful and even may degrade the decisions made by the model. Then the selected feature set is normalized between the range zero and one using the Min-Max scaling technique. Next, we chunked the collected data series into t length windows. Then the input to our AI algorithm is a sequence of sensor readings, S, of length t: $S = \{(s_{11}, s_{21}, \cdots, s_{k1}), \cdots, (s_{1t}, s_{2t}, \cdots, s_{kt})\}$.

In this study, we used a Deep Learning based approach to predict human activity given the sensor data. Hence, the final module comprises a stacked Gated recurrent unit (stacked-GRU) which works as the feature extractor. The stacked GRU consists of two GRU layers. The first layer has t number of cells and the second layer has r number of cells. t and r are hyperparameters. The input is given to the first layer and 2nd GRU layer output to a fully connected layer after a dropout layer to reduce the overfitting.

The Rectified Linear (ReLU) is used as the activation function of this fully connected layer. The output of this dense layer is then input into the last dense layer with a SoftMax activation function which returns the probability for each output class. The output class with the highest probability is taken as the prediction class. We use the categorical cross-entropy loss for the classification loss.

4.2 Proposed explainable human activity recognition model

We extended the previously introduced HAR system to generate explanations in addition to the activity label. For this purpose, we used SHAP [21], which is a model-agnostic post-hoc explanation method that calculates Shapley values to understand the importance of each feature used by a black-box AI model toward prediction. Since our input to the model is a sequence of k features in t time-steps, we get $k \times t$ SHAP values for a particular input sample. Nevertheless, SHAP values are numerical and are complex to understand by lay users (e.g., the caregivers).

As a solution, we simplified SHAP explanations as shown in Fig. 6. This was done by converting them into linguistic explanations as shown

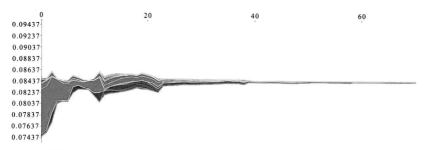

Fig. 6 SHAP explanation for a sample activity.

in Fig. 5 part (v). First, we calculated the summation of SHAP values over t timesteps and select the n features with the highest SHAP values as the most influencing features. This set includes the n features that caused the AI algorithm to predict the activity class. We set n to 3 in our experimental study. Next, the statistics such as variance, minimum, maximum, number of peaks, and periodicity of these n features are calculated.

These statistics indicate the discriminative characteristics of sensor signals that influenced the AI algorithm decision. Finally, we use a rule-based algorithm employing the calculated statistics to describe the characteristics of time-series input data to generate linguistic explanations. We present the linguistic explanation along with the graph showing the n most influencing features as the final explanation as shown in the last process of Fig. 5. The details are described later in this section with experimental data as shown in Fig. 7.

4.3 Experimental study

4.3.1 Dataset

The MHealth dataset comprises recordings of 3 wearable sensors for 10 participants performing 12 typical daily activities [39]. These include standing still, sitting, and relaxing, lying down, walking, climbing stairs, waist bends forward, frontal elevation of arms, knees bending, cycling, jogging, running, and jumping front and back. Three sensors have been placed on the body at the chest, left ankle, and right wrist to record these activities. Sensors placed on the left ankle and right wrist are used to take the readings of the embedded triple-axis accelerometer, gyroscope, and magnetometer readings while performing these activities. The sensor placed on the chest area has 2 embedded sensors, a tri-axis accelerometer to measure the acceleration, and 2 ECG

Predicted activity - Walking
The system identifies (a) as "walking" because
    All 3 sensor readings are periodic and aLeftAnkleZ has high
variation while aRightLowerArmY and aLeftAnkleY have medium

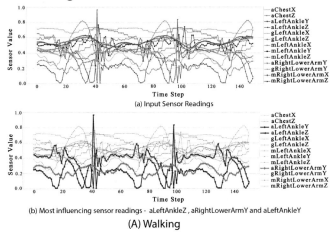

(A) Walking

Predicted activity - Waist bends forward
The system identifies (a) as "Waist bends forward" because
    aRightLowerArmY and aChestX are aperiodic and have medium
variation. Also, aLeftAnkleZ is flat. aRightLowerArmY and aLeftAnkleY
have medium variations.

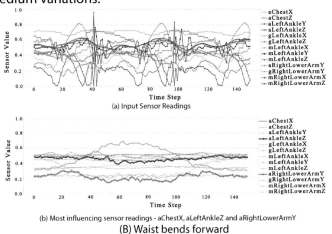

(B) Waist bends forward

Fig. 7 Explanations generated by the proposed HAR System for (A) Walking and (B) Waist bends forward.

leads which can be used to measure the basic heart rate. Altogether there are 23 distinct features recorded at each time step along with the activity label. Observations were recorded at 50 Hz. The MHealth contains 344,116 samples altogether. We use 70% of random samples as the training data and the remaining 30% as the testing data.

4.3.2 Data preprocessing

The first preprocessing step was feature selection. Using the mutual information feature selection technique, chose the top 13 features that have a strong correlation with the activity labels. The data is then normalized using min-max normalization. Finally, we considered a sliding window of 3 s with 50% overlap to set the number of the timesteps, so that there were 150 time-steps for each input. The dimension of the input to our AI model is $b \times 13 \times 150$, where b is the batch size.

4.3.3 Human activity recognition model

Recall that we use a stacked GRU as the HAR model consists of two GRU layers. The first layer has 150 cells and the number of cells in the second layer is set to 100. The model is trained for 20 epochs while optimized using the Adam optimizer with a learning rate of 0.01. The dropout rate is set to 0.5. The accuracy of the proposed deep learning model is calculated as 93% on the recorded MHealth dataset.

4.3.4 Generated multimodal explanations for human activity recognition

This section presents sample multimodal explanations generated by the proposed approach for three activity instances, lying down, walking, and waist bends forward. For each sample, we presented the raw input, the most influential three features, and the linguistic explanation. Each sensor reading is denoted by an identification name. This name is formally defined as $\ll F \gg \ll sensorLocation \gg \ll axis \gg$ Where, $F = \{a, g, m\}$, $sensorLocation = \{LeftAnkle, RightAnkle, RightLowerArm, LeftLowerArm, Chest\}$ and $axis = \{x, y, z\}$. The symbol "a" represents the accelerometer, "g" represents the gyroscope, and "m" represents the magnetometer.

Fig. 7 Shows the explanations generated by the proposed HAR System for walking and waist bends forward. For the activity predicted as walking, the most impacting three features are $aLeftAnkleZ$, $aRightLowerARmY$ and $aLeftAnkleY$. The explanation describes that the prediction was caused by the periodicity of all three signals and the high variation of $aLeftAnkleZ$.

Generally walking is a periodic task and it aligns with the explanation also. For the activity predicted as "waist bend forward," the most important features are $aRightLowerArmY$, $aChestX$, and $aLeftAnkleZ$. In a waist bend, the arm and the chest move, but that is not a periodic activity, and the leg stays stationary which is also conveyed by the explanation.

With these detailed multimodal explanations, the user gets an understanding of the reasons about a particular set of sensor readings is predicted to be a certain activity by the system. Further, these explanations are well aligned with the domain knowledge. Therefore, the users can keep trust in the predictions the system makes. We evaluated whether explanations help to improve user trust in the system in the next section using a user study.

5. User attitude toward adoption of explainable human activity recognition systems

This section presents the design and analysis of the user perception of XAI in ELE. Our model is based on the Technology Acceptance Model (TAM) [40], which measures users' attitudes toward the adoption of a Human Activity Recognition System providing explanations along with the decisions. We have conducted another post-survey for this purpose. Forty volunteers participated in this survey. We demonstrate that explanations improve user trust in the system which eventually leads to increased adoption of HAR systems.

5.1 User perception study design

The TAM explains the determinants of computer acceptance. Original TAM and its variations have been extensively applied to evaluate the user's attitude toward various computer-based systems. TAM is based on two core concepts: perceived usefulness and perceived ease of use. Furthermore, the TAM has been derived from the Theory of Reasoned Action, which states that individuals adopt a specific behavior if they perceive it leads to a positive outcome. We argue that we can adapt TAM to evaluate the users' intention to accept an AI-based ELE system if they can understand the system and if they can trust the system. Hence, we set the determinants of system acceptance as perceived understandability and perceived trust in the system. These two determinants are influenced by the individual characteristics of the users and system characteristics. The proposed conceptual model is shown in Fig. 8.

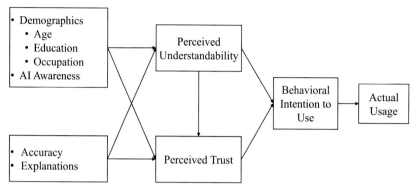

Fig. 8 Conceptual model for validation survey.

The survey consisted of four sections. In the first part, we collected demographic information from the respondents. The second section consists of a questionnaire to determine the AI awareness of the participants.

The following hypotheses were used to examine whether the use of XAI in an ELE system improves user trust and usability of the system.

- **Ha:** Understandability has an indirect positive link with behavioral intention
- **Hb**: Perceived trust has a positive relationship with behavioral intention.
- **Hc:** Understandability has a direct relationship with behavioral intention.
- **Hd:** Understandability highly influences the perceived trust of the system.
- **He:** Understandability and perceived trust in the system have a strong positive relationship to behavioral intention.
- **Hf:** Behavioral intention has a strong positive relationship to system usage.
- **Hg:** Perceived importance of XAI and perceived trust in the system have a strong positive relationship to actual usage.

Next, the participants were given an activity to evaluate whether the presence of explanations enhances the users' understandability of the system. For that, they were first given input sensor data and the HAR system's prediction for a set of activity samples. Then, given those examples, users were asked to mimic the HAR system and guess the system's forecast for a set of new activity instances. Next, participants were again presented with the first set of activity instances, but in addition to the prediction, we provided

Table 7 Construct items defined to test the hypotheses.

Perceived understandability of the system
1. Assume that an AI system makes decisions for you in your day-to-day activities. Is it to know the reasons behind such decisions?
2. Explanations helped me to understand the reasons for the AI system's decisions.

Perceived trust in the system
1. Explanations improved my trust in the AI system.
2. I am inclined to trust the AI system that gives explanations along with the decisions over the AI system that only gives decisions.

Behavioral intention to use
1. I think that using the system giving explanations is a good idea.
2. I think that using the system giving explanations is beneficial for me.
3. I have a positive perception of using the system that gives explanations.

Usage
1. I am inclined to trust the AI system that gives explanations along with the decisions over the AI system that only gives decisions.
2. I would like to use the AI system with explanations

multimodal explanations for the prediction. Then, another set of activity instances was given and asked the participants to guess the system's prediction for those samples. In the final section, a questionnaire based on the constructs shown in Table 7 was used to collect the participants' attitudes regarding the understandability, trust and intention to use the system. The questionnaire consisted of Likert scales ranging from strongly disagree to strongly agree.

5.2 Analysis of user attitude toward HAR with explanations

In order to test the proposed research model, a post-survey was conducted in August 2022. The total number of participants in the study was 40, out of which 25 (62.5%) were male and 15 (37.5%) were female. The participants' mean age was 28.62 years with a standard deviation of 3.72. The survey analysis was done based on Correlation analysis and regression analysis. The correlation analysis results are given in Table 8, which shows a high correlation between each pair of constructs.

Regression analysis between the constructs was used to measure the support for each hypothesis at a significance level of 95%. As shown in Table 9,

Table 8 Correlation of constructs.

Constructs	(1)	(2)	(3)	(4)
(1) Perceived understandability	1	0.5460	0.6674	0.5038
(2) Perceived trust	0.5460	1	0.6321	0.3553
(3) Behavioral intention to use	0.6674	0.6321	1	0.5543
(4) Usage	0.5038	0.3553	0.5543	1

Table 9 Hypotheses test results.

Hypothesis	P value
Ha: Perceived understandability has an indirect positive relationship with behavioral intention	0.4435
Hb: Perceived trust has a positive relationship with behavioral intention	1.21E-05
Hc: Perceived understandability has a direct relationship with behavioral intention	2.55E-06
Hd: Perceived understandability significantly influences the perceived trust in the system	0.000268
He: Perceived understandability and perceived trust in the system have a strong positive relationship to behavioral intention	4.24E-07
Hf: Behavioral intention has a high positive relationship with system usage	0.000206
Hg: Perceived understandability and perceived trust in the system have a strong positive relationship to actual usage	0.003529

the hypotheses Hb, Hc, Hd, He, Hf, and Hg are supported, and Ha is not. Further, Fig. 9 shows respondents' prediction score distribution before and after seeing the explanations. It can be seen that participants were able to correctly predict the system behavior after seeing the explanations. The participants' average prediction scores before and after seeing the explanations are 6.4 and 7.1, respectively, showing that their understanding of the system improved after exposure to the explanations. The above results indicate strong evidence that the explanations given by the proposed explanation generation mechanism support the behavioral intention and, ultimately, the usage of HAR AI systems in IoT environments.

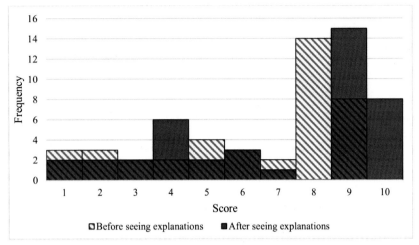

Fig. 9 The distribution of the number of correct predictions by the participants before and after seeing the explanations.

6. Discussion

6.1 Lessons learned

In this study, we explored the user perception and requirements of XAI. In addition, the improvement of user trust in the AI-based application and the increase in the adoption of these applications, with the exposure to explanations were analyzed. The main lessons learned are discussed as follows.

(1) *Being aware of the rationale of AI systems is vital*: Over 85% of our survey participants have indicated that knowing the reasons for AI decisions is essential for them. Our data analysis reveals that this perception is consistent across different demographics. However, the perceived importance of XAI depends on the AI awareness of the participant. Participants with high AI awareness found it more important to access the explanations. Recall that the participants in our survey represented various demographic groups and might not have any expertise in AI. Further, most of the systems in ELEs directly interact with lay users who may also not have any expertise in AI, such as healthcare workers remotely monitoring patients. Hence, these findings emphasize the importance of integrating explainability into ELE applications.

(2) *Explanations should be easy to understand, faithful, and interactive*: Our literature review revealed that existing work incorporating XAI into

ELE applications had merely used current XAI methods without customizing them for the specific requirements of the ELEs. However, explanations generated by some of the widely used existing XAI methods are not understandable even to data scientists [27]. Hence, the explanations provided for the AI decisions of ELE systems should cater to these users.

Participants of our survey expected different requirements from the explanations. The most highly rated requirement is being easy to understand. This is supported by the fact that most participants preferred linguistics or prototype-based explanations over feature attributions. Hence, we should adapt concept-based explanations for ELE applications. However, most of the existing concept-based explanation mechanisms focus on image data, whereas most ELE applications rely on tabular data, time-series data, or multimodal data. Therefore, more work needs to be done on generating concept-based explanations for the data modalities commonly used in ELE applications.

The second most desirable characteristic of explanations is being faithful; explanations should reveal the actual reasons that caused the AI decisions. Since some of the ELE applications are high stakes, the user must know the system's true rationale. However, studies have demonstrated that some of the existing feature attribution methods are not faithful to the underline system [41]. Hence, when adapting a current XAI method, ELE researchers should be cautious about the fidelity of the explanations generated by the said method. Further, developing faithful explanations should be explored more in the XAI community.

Another desirable characteristic of explanations, as indicated by our survey participants, is being interactive. Users should be presented with explanations only when they ask for them, able to change the granularity (from global explanations to local explanations and vice versa) and change the modality of explanations (visual or linguistic) for their preference. In addition to these characteristics, the participants have indicated that they prefer explanations to be descriptive and consistent.

(3) *Explanations are crucial for high-stake ELE applications*: Our survey revealed that users expect high-stake ELE applications with crucial impacts such as healthcare or security to be accompanied by explanations. Hence, researchers developing ELE applications for these domains must invest extra effort to provide explanations for the AI decisions made by such systems.

(4) Explanations lead to an increase in the chance of adopting the system: The findings of our second survey endorse that having access to explanations improves user trust in the ELE application leading to higher adoption of those systems ($P = 0.0035$). Hence, we can conclude that explanations play a critical role in increasing the adoption of ELE applications in the real world.

6.2 Contribution toward ELE

The recent interest in ELE has created continual momentum in relevant research studies to incorporate novel and next-generation technologies to realize better ELEs. The state-of-the-art ELE applications increasingly depend on AI approaches, tools, and techniques to provide enhanced performance. Smart characteristics of ELE are primarily achieved through the automation and interfacing of physical space through IoT. With continual listening and communication among the swarm of sensors and actuators, these implementations generate an enormous amount of raw data points, which can be used to develop AI models. Once deployed, these AI models can take over the intelligent automation of the ELE and further enhance their accuracy and responsiveness through a continuous feedback cycle. Accordingly, a higher degree of trustworthiness of the deployed AI model is expected, assuring that the AI models manipulate the environment through IoT in the intended way.

Nevertheless, the decision-making process of the existing AI models is opaque, reducing the trustworthiness of AI-facilitated ELE. The lack of end-user trust in ELE applications leads to fewer adaptions of those systems in real-world applications, especially in mission-critical domains such as healthcare. Yet, most of the existing studies in IoT for ELEs pay less attention to enhancing the transparency of ELEs and establishing end-user trust in these applications.

The study conducted in this chapter pointed out the pressing necessity of adapting XAI techniques to ELE applications and the users' expectations of the explanations for the decisions made in those applications. This transparency enables identifying potential errors or biases in the AI-driven ELE that could impact the decisions the users are subjected to. Overall, XAI can improve the transparency and trustworthiness of AI-driven ELE, allowing users to make more informed decisions and have greater control over their living environments. Hence, this chapter brings a new research direction to the attention of the ELE research community that is paramount for developing and adapting ELE applications in the real world. Furthermore, the findings of this chapter can be used as a guideline for developing trustworthy ELE applications in the future.

7. Conclusion

This chapter presents an in-depth analysis of the user perception of the importance of XAI, the desirable characteristics of explanations, and the enhancement of the adaptability of ELE applications with exposure to understanding explanations. We first conducted an online user survey with the participation of 326 volunteers to understand the user perception of XAI and their requirements for explanations. The study reveals that irrespective of the demographic differences, most participants find that being aware of the rationale of the AI system's decision is vital. Moreover, the participants expected that explanations should be easy to understand, faithful, and interactive, among other desirable characteristics. They have also indicated that they prefer concept-based explanations over feature attributions. Based on the insights gathered from this survey, we proposed a new multimodal explanation method to explain the decisions of a sensor-based HAR model. The proposed method can explain HAR model decisions using visual and linguistic explanations. Finally, we conducted another user survey for the proposed conceptual model based on the TAM model. It assessed whether the users' understandability and trust in AI-based applications improve and affect the adoption of the system if they are presented with the rationale of the decisions made by the system. The data analysis reveals that explanations enhance the user's understandability and trust in the system. Further, we observed that perceived understandability and trust strongly impact the intention to use the system and actual usage. The findings of this research are valuable for both the XAI community and the ELE community. It is suggested to invest more effort in developing methods that generate easy-to-understand and faithful explanations, for the XAI community. Furthermore, it is recommended to integrate explanations into the AI-based ELE applications to increase the adoption of those systems in the real world.

References

[1] S. Oniani, S. Mukhashavria, G. Marques, V. Shalikiani, I. Mosashvili, Mobile computing technologies for Enhanced Living Environments: a literature review, in: The Big Data-Driven Digital Economy: Artificial and Computational Intelligence, 2021, pp. 21–32.

[2] J. Shin, Y. Park, D. Lee, Who will be smart home users? An analysis of adoption and diffusion of smart homes, Technol. Forecast. Soc. Change 134 (2018) 246–253.

[3] C. Liang, D. Liu, L. Qi, L. Guan, Multi-modal human action recognition with sub-action exploiting and class-privacy preserved collaborative representation learning, IEEE Access 8 (2020) 39920–39933.

[4] D. Meedeniya, Deep Learning: A Beginners' Guide, CRC Press, 2023.

[5] R. Caruana, Y. Lou, J. Gehrke, P. Koch, M. Sturm, N. Elhadad, Intelligible models for healthcare: predicting pneumonia risk and hospital 30-day readmission, in: *Proc. of the 21st ACM SIGKDD International Conference on Knowledge Discovery and Data Mining*, 2015, pp. 1721–1730.

[6] S. Dasanayaka, V. Shantha, S. Anupa Silva, D. Meedeniya, T. Ambegoda, Interpretable machine learning for brain tumour analysis using MRI and whole slide images, Softw. Impacts 13 (2022) 100340.

[7] L.A. Hendricks, Z. Akata, M. Rohrbach, J. Donahue, B. Schiele, T. Darrell, Generating visual explanations, in: Proc. of the European Conference on Computer Vision, Amsterdam, The Netherlands, 2016, pp. 3–19.

[8] R.A. Khalil, N. Saeed, M. Masood, Y.M. Fard, M.-S. Alouini, T.Y. Al-Naffouri, Deep learning in the industrial internet of things: potentials, challenges, and emerging applications, IEEE Internet Things J. 8 (14) (2021) 11016–11040.

[9] S. Senarath, P. Pathirana, D. Meedeniya, S. Jayarathna, Customer gaze estimation in retail using deep learning, IEEE Access 10 (2022) 64904–64919.

[10] H. Padmasiri, R. Madurawe, C. Abeysinghe, D. Meedeniya, Automated vehicle parking occupancy detection in real-time, in: *Moratuwa Engineering Research Conference (MERCon)*, 2020, pp. 644–649.

[11] H. Padmasiri, J. Shashirangana, D. Meedeniya, O. Rana, C. Perera, Automated license plate recognition for resource-constrained environments, Sensors 22 (4) (2022) 1434.

[12] D. Meedeniya, H. Kumarasinghe, S. Kolonne, C. Fernando, I. De la Torre Díez, G. Marques, Chest X-ray analysis empowered with deep learning: a systematic review, Appl. Soft Comput. 126 (2022) 109319.

[13] M.A. Gul, M.H. Yousaf, S. Nawaz, Z. Ur Rehman, H. Kim, Patient monitoring by abnormal human activity recognition based on CNN architecture, Electronics 9 (12) (2020) 1993.

[14] M.Z. Uddin, A. Soylu, Human activity recognition using wearable sensors, discriminant analysis, and long short-term memory-based neural structured learning, Sci. Rep. 11 (1) (2021) 16455.

[15] M. Sundararajan, A. Taly, Q. Yan, Axiomatic attribution for deep networks, in: Proc. of the 34th International Conference on Machine Learning, Sydney, Australia, vol. 70, 2017, pp. 3319–3328.

[16] B. Kim, et al., Interpretability beyond feature attribution: quantitative testing with concept activation vectors, in: Proc. of the 35th International Conference on Machine Learning, Stockholm, Sweden, 2018, pp. 2668–2677.

[17] A. Ghorbani, J. Wexler, J.Y. Zou, B. Kim, Towards automatic concept-based explanations, in: *Proc. of the 33rd Conference on Neural Information Processing Systems* (NeurIPS), Vancouver, Canada, 2019, pp. 1–10.

[18] S. Wickramanayake, W. Hsu, M.L. Lee, FLEX: faithful linguistic explanations for neural net based model decisions, in: Proc. of the AAAI Conference on Artificial Intelligence, Hawaii, USA, vol. 33, 2019, pp. 2539–2546.

[19] C. Chen, O. Li, D. Tao, A. Barnett, C. Rudin, J.K. Su, This looks like that: deep learning for interpretable image recognition, in: *Advances in Neural Information Processing Systems*, Vancouver, Cananda, 32, 2019.

[20] T.-T.-H. Le, H. Kim, H. Kang, H. Kim, Classification and explanation for intrusion detection system based on ensemble trees and SHAP method, Sensors 22 (3) (2022) 1154.

[21] S.M. Lundberg, S.-I. Lee, A unified approach to interpreting model predictions, in: *Proc. of the Advances in Neural Information Processing Systems*, Long Beach, USA, vol. 30, 2017, pp. 4768–4777.

[22] M.P.S. Lorente, E.M. Lopez, L.A. Florez, A.L. Espino, J.A.I. Martínez, A.S. de Miguel, Explaining deep learning-based driver models, Appl. Sci. 11 (8) (2021) 3321.

[23] J. Pamela S, et al., Autism Spectrum disorder prediction by an explainable deep learning approach, Comput. Mater. Contin. 71 (1) (2022) 1459–1471.

[24] L. Arrotta, G. Civitarese, C. Bettini, DeXAR: deep explainable sensor-based activity recognition in smart-home environments, J. ACM 6 (1) (2022) 1–30.

[25] S. Dasanayaka, S. Silva, V. Shantha, D. Meedeniya, T. Ambegoda, Interpretable machine learning for brain tumor analysis using MRI, in: *Proc. of the 2nd International Conference on Advanced Research in Computing (ICARC)*, 2022, pp. 212–217.

[26] D. Thakker, B.K. Mishra, A. Abdullatif, S. Mazumdar, S. Simpson, Explainable artificial intelligence for developing smart cities solutions, Smart Cities 3 (4) (2020) 1353–1382.

[27] H. Kaur, H. Nori, S. Jenkins, R. Caruana, H. Wallach, J. Wortman Vaughan, Interpreting interpretability: understanding data Scientists' use of interpretability tools for machine learning, in: CHI Conference on Human Factors in Computing Systems, Honolulu, USA, 2020, pp. 1–14.

[28] M. Langer, et al., What do we want from Explainable Artificial Intelligence (XAI)? Artif. Intell. 296 (2021) 103473.

[29] M. Riveiro, S. Thill, 'That's (not) the output I expected!' On the role of end user expectations in creating explanations of AI systems, Artif. Intell. 298 (2021) 103507.

[30] D. Wang, Q. Yang, A. Abdul, B.Y. Lim, Designing theory-driven user-centric explainable AI, in: Proc. of the CHI Conference on Human Factors in Computing Systems, Glasgow, UK, 2019, pp. 1–15.

[31] P. Langley, B. Meadows, M. Sridharan, D. Choi, Explainable agency for intelligent autonomous systems, in: Presented at the Thirty-First AAAI Conference on Artificial Intelligence, San Francisco, USA, 2017.

[32] D. Alvarez Melis, T. Jaakkola, Towards robust interpretability with self-explaining neural networks, in: *Advances in Neural Information Processing Systems*, Montréal, Canada, vol. 31, 2018.

[33] M.T. Ribeiro, S. Singh, C. Guestrin, 'Why should i trust you?': Explaining the predictions of any classifier, in: *Proc. of the 22nd ACM SIGKDD Int. Conference on Knowledge Discovery and Data Mining*, 2018, pp. 1135–1144.

[34] R.R. Selvaraju, M. Cogswell, A. Das, R. Vedantam, D. Parikh, D. Batra, Grad-CAM: visual explanations from deep networks via gradient-based localization, in: Proc. of the IEEE International Conference on Computer Vision, Venice, Italy, 2017, pp. 618–626.

[35] C.B. Hilton, et al., Personalized predictions of patient outcomes during and after hospitalization using artificial intelligence, NPJ Digit. Med. 3 (51) (2020) 1–8.

[36] M.T. Ribeiro, S. Singh, C. Guestrin, Anchors: high-precision model-agnostic explanations, in: Proc. of the AAAI Conference on Artificial Intelligence, Louisiana, USA, 2018, pp. 1527–1535.

[37] M. Hou, L. Wu, E. Chen, Z. Li, V.W. Zheng, Q. Liu, Explainable fashion recommendation: a semantic attribute region guided approach, in: Proc. of the Twenty-Eighth International Joint Conference on Artificial Intelligence (IJCAI-19), Macao, China, 2019, pp. 4681–4688.
[38] X. Chen, R. Mottaghi, X. Liu, S. Fidler, R. Urtasun, A. Yuille, Detect what you can: detecting and representing objects using holistic models and body parts, in: Proc. of the IEEE Conference on Computer Vision and Pattern Recognition, Columbus, OH, USA, 2014, pp. 1971–1978.
[39] MHEALTH dataset, (2022) UC Irvine Machine Learning, [Online: https://archive-beta.ics.uci.edu/ml/datasets/mhealth+dataset], Accessed (May 2022).
[40] F.D. Davis, V. Venkatesh, A critical assessment of potential measurement biases in the technology acceptance model: three experiments, Int. J. Hum. Comput. Stud. 45 (1996) 19–45.
[41] J. Adebayo, J. Gilmer, M. Muelly, I. Goodfellow, M. Hardt, B. Kim, Sanity checks for saliency maps, in: *Proc. of the 32nd International Conference on Neural Information Processing Systems*, 2018, pp. 9525–9536.

About the authors

Sandareka Wickramanayake is a senior lecturer at the Department of Computer Science & Engineering, University of Moratuwa, Sri Lanka. She obtained her PhD from the National University of Singapore in 2022, where she was supervised by Prof. Wynne Hsu and Prof. Mong Li Lee. She earned her BSc (Hons.) in Computer Science and Engineering from the University of Moratuwa in 2014. She joined the University of Moratuwa as a lecturer in 2015. Dr. Sandareka's interests lie in Interpretable Machine Learning, mainly designing accurate and interpretable Deep Neural Networks (DNNs) and applications of DNNs for real-world problems. She has multiple publications in top peer-reviewed conferences such as AAAI and NeurIPS.

Sanka Rasnayaka is a lecturer at the School of Computing, National University of Singapore. He received a BSc degree (Hons.) in computer science and engineering from the Engineering Faculty, University of Moratuwa, Sri Lanka, in 2016 and a PhD degree in computer science from the School of Computing, National University of Singapore, for his work on continuous authentication for mobile devices, in 2021. His research interests include applications of AI and computer vision. He mainly works in the fields of biometrics and authentication.

Madushika Gamage is a post graduate student currently pursuing a master's in computer science at the University of Moratuwa, Sri Lanka, following her bachelor's degree in Electronics and Telecommunications from the same esteemed institution. She had been working in leading software companies for few years in Sri Lanka as well as in Singapore. Her research interests are in AI, Deep Learning, Explainable AI, Human-computer Interaction.

Dulani Meedeniya is a professor in Computer Science and Engineering at the University of Moratuwa, Sri Lanka. She holds a PhD in Computer Science from the University of St Andrews, United Kingdom. She is the director of the Bio-Health Informatics group in her department and engages in many collaborative research. She is a co-author of 100+ publications in indexed journals, peer-reviewed conferences and book chapters. She serves as a reviewer, program committee and editorial team member in many international conferences and journals. Her main research interests are software modelling and design, Bio-Health Informatics, Deep Learning and Technology-enhanced learning. She is a Fellow of HEA (UK), MIET, Senior member of IEEE, Member of ACM and a Chartered Engineer registered at EC (UK).

Indika Perera is a professor in Computer Science and Engineering at the University of Moratuwa, Sri Lanka. He holds a PhD (St Andrews, UK) MBS (Colombo), MSc (Moratuwa), PGDBM (Colombo) and BSc Eng. (Hons.) (Moratuwa). He is a Fellow of HEA (UK), MIET, SMIEEE and a Chartered Engineer registered at EC (UK) and IE (SL). His research interests are Software Engineering, Intelligent Systems, Technology Enhanced Learning, User Experience and Immersive Environments.

CHAPTER TWO

Human behavioral anomaly pattern mining within an IoT environment: An exploratory study

Rosario Sánchez-García[a], Alejandro Dominguez-Rodriguez[b,c], Violeta Ocegueda-Miramontes[a], Leocundo Aguilar[a], Antonio Rodríguez-Díaz[a], Sergio Cervera-Torres[d], and Mauricio A. Sanchez[a]

[a]Universidad Autónoma de Baja California, Baja California, Mexico
[b]Psychology, Health, and Technology, University of Twente, Enschede, The Netherlands
[c]Health Sciences Area, Universidad Internacional de Valencia, Valencia, Spain
[d]Multimodal Interaction Lab. Leibniz–Institut für Wissensmedien (IWM), Tübengen, Germany

Contents

Abstract

A Psychological assessment is fundamental for the detection of clinical mental disorders. However, standard psychometric tools such as questionnaires, which are the gold standard for assessing clinical disorders, face important drawbacks regarding subjective bias. Accordingly, new methods and technologies are coming to the fore to complement psychometric assessments so that more consistent and replicable

Advances in Computers, Volume 133
ISSN 0065-2458
https://doi.org/10.1016/bs.adcom.2023.10.003

assessments can potentiate patient-focused diagnosis accuracy. In this sense, the development of Internet of Things (IoT) networks has been part of the technological advances that are characteristic of Industry 4.0, due to the large amount of information provided by networked sensors regarding the environment and the interaction of individuals in it, allowing the detection of behavioral patterns exercised. This paper proposes a data analysis of human behavioral patterns from a connected home environment. The potentials of pattern mining techniques are investigated for detecting behavioral anomalies within such patterns. Results show that detecting anomalies within human behavioral patterns is possible. We argue that with such promising results, a system could be potentially applied under contexts such as suicide prevention or discovering other undiagnosed mental disorders that individuals may present throughout their life.

1. Introduction

Appropriate psychological assessment of individuals for clinical purposes is fundamental to determine the presence or absence of mental disorders classified in the Diagnostic and Statistical Manual of Mental Disorders, 5th edition (DSM-5) of the American Psychiatric Association [1]. However, due to its young and recent history, there is still disagreement among experts regarding the progress on psychometric measurements, even in fundamental concepts such as validity. As [2], indicates, "*validity is widely acknowledged to be the most fundamental consideration in the development of psychometric measures.… but there are a surprising number of conceptualizations of validity, some not even closely analogous*" [2].

In the light of such limitations, novel methodologies to potentiate objective psychological assessments and, in turn, more reliable diagnosis are coming to the fore. Concretely, technological developments such as biosensors and internet-connected devices such as smartphones are being exploited to monitor and gather biometric data, which are subsequently submitted to sophisticated data analyses such as Machine Learning (ML) techniques [3]. ML is a branch of artificial intelligence (AI), which relies on the notion that systems can learn from the data and make automated decisions with minimal human intervention [4].

The authors of [5] used intelligent data mining based on a genetic ML algorithm as a more sophisticated methodological approach to assess and provide a more fine-grained and objective mental health diagnosis. Research has also been conducted on suicide prevention through the use of AI. A systematic review was also provided by [6] that included 87 studies on social networks, which concluded that AI and ML had an accuracy higher than 90% in predicting suicidal behaviors. Moreover, ML methods are being applied to advance theory and assessment on personality. In particular, novel research trends focus on personality assessment in social media and digital records. An example was published on the web page of an article of the American Psychological Association [7], that presents the results of [8], where the researchers attempt to predict the Big Five personality dimensions [9], using as a primary source of information the activity of the users on their cellphones, a standardized personality assessment questionnaire, and an ML algorithm. The researchers observed that communication, social-related behaviors, and app use patterns, were important personality predictors.

However, ML is a relatively recent methodological approach used in Psychology that needs to expand to personality assessment by embedding it in a more comprehensive construct validation framework. For example, in the study by [10], the authors identify three generations of studies using ML in personality assessment. The first generation focused on introducing ML for personality assessment, such as applying text analysis tools to the language written in the personal profiles and messages [10,11]. The second generation included studies that gathered and analyzed broad amounts of data from users of social networks, primarily Facebook, with the purpose of optimizing and predicting the validity of these approaches. The third generation focused on analyzing whether ML personality assessment could improve the traditional assessment methods in Psychology, such as self-and-peer reports.

Going further, to analyze the mental health state of the individual through ML, is using bodily and behavioral information. As indicated by [12], "*The face and the body both normally contribute in conveying the emotional state of the individual*" [12]. Other authors in the field of ML and emotions also identify that body language has a critical role regarding the communication process and provides cues that allow the detection of various aspects of an individual's mental state [13]. Research on autistic spectrum disorder (ASD) is a good reference. For instance, eye-tracking techniques, body motion devices, and wearable sensors (e.g., wristbands to measure electrodermal activity) are being used to gather large amounts of data to predict, by ML and computational models, the ADS diagnosis of children [14].

However, analyzing large amounts of behavioral data creates new questions about how to find and recognize meaningful patterns of information.

According to [15] *"Pattern Mining can be defined as the nontrivial extraction of implicit, previously unknown, and potentially useful information, in the form of frequent patterns, from data"*. In this regard, the development of technologies for the detection of anomalies through an established pattern of behaviors is of interest in multiple areas of computer science, and based on recent controversies regarding the level of replicability of behavioral research by using statistical inference, the development of more efficient techniques to analyze the results of psychological experiments have been proposed [16].

Nowadays, the field of Psychology is exploring the possibility of relying on these technologies that characterize Industry 4.0, which involve a much more ubiquitous and mobile internet, more powerful sensors, and the application of AI due to the impressive advances achieved by AI-driven by exponential increases in computing and the vast information that can be obtained through environments related to IoT. An IoT environment permits the acquisition of large amounts of information about the person who interacts within it, and thanks to the cost reduction in the sensors and the increase in their performance, their implementation in diverse applications has become more practical and feasible [17].

Complementing IoT environments with disciplines such as pattern mining and psychology would support an essential objective in the research of intelligent spaces, which is to detect and predict behaviors [18] based on data generated by users and detected by the environment, even generating information to warn of anomalous behaviors that could occur at some point in the individual [19,20]. Therefore, while appropriate psychological measures improve, new ways to evaluate and/or identify the mental health states of the individuals should be explored. It is highly relevant to expand the scientific evidence about the evaluation of psychological states not only measured by traditional methods in Psychology, such as psychometrics, but by a collaborative system.

In this chapter, the main contributions are as follows: (1) an IoT environment is designed with sensors selected for significant activity monitoring related to behavioral patterns based on general psychological analysis of relevant behaviors; (2) a data arrangement process that takes the monitored data from the IoT environment in preparation for the data mining process; (3) and a data mining process which obtains different clusters of general behavior as well as anomalous behaviors for the individual being monitored. The objective of this study is to explore the viability of detecting anomalous behaviors in individuals based solely on their interactions with sensors in their home environment as well as other monitoring technologies.

This document is divided into the following sections: Section 2 details the IoT sensor network, data description, and data preparation. Section 3 describes the methodology. Section 4 shows a discussion of results. And Section 5 provides concluding remarks on the proposed approach.

2. Materials and methods

This study was designed and conducted by a multidisciplinary team involving researchers in the areas of computer engineering, computer science, and clinical psychology.

2.1 Participant

The following requirements were needed to be fulfilled by the candidate to be able to participate in the study: (1) not having a motor disability, (2) not consuming drugs, (3) not consuming alcohol, (4) individual housing, (5) legal adulthood age (18+), (6) not having pets, and (7) to live alone.

The recruited participant was a 25-year-old male, living alone in Tijuana (Mexico). The participant was informed of the procedures to follow, which devices were to be implemented in their environment and where the smart tools that he should carry (commercial smartwatch band and cell phone) for daily monitoring. An explanation was also given of privacy policies under which this study works.

After signing an informed consent form explaining the procedures, an extensive psychological evaluation was carried out with validated psychometric instruments to screen for affective and mental disorders. A trained clinical psychologist applied the instruments to assess the mental health of the participant. Although not a necessity for the success of the experimentation, the proposed approach and methodology should, in any case, work on discovering behavioral patterns as well as anomalous conduct in any person. It was decided to be best if a participant with no apparent symptoms of mental health issues was initially included in the experiment, in order to reduce any unforeseen issues. Sixteen psychological instruments were used.

2.2 IoT environment design

The general IoT sensor network was designed in-house to enable high control over the flow of information related to target conducts, which an

Table 1 Used sensors and locations for significant conduct measurement.

Area	Sensor type	Location	Detected activity	Measured conduct
Living room	PIR + LDR	Entrance	Motion	Movement Sedentarism
Kitchen	PIR + LDR	Above table	Motion	Eating Sedentarism
	Ultrasonic + Switch	Stove Pantry drawer	Stove use Open/Close	Cooking Eating
	Switch	Refrigerator	Open/Close	Cooking Eating
Dormitory	PIR + LDR	Entrance	Motion	Movement Sedentarism
	AC Meter	Laptop charger	Use of device	Recreation
	Flex Sensor	Bed	Motion in bed	Sleep
Lavatory	PIR + Humidity	Entrance	Motion Use of shower	Take shower

expert on clinical psychology recommended as relevant to the experiment. Table 1 shows all selected areas and the used sensors in relation to detected activities and measured behaviors.

PIR (HC-SR501) is a passive infrared sensor for motion detection. LDR is a light-dependent resistor for measuring light intensity. Ultrasonic (HC-SR04) is a distance measurement sensor. Switch is a magnetic reed switch for detecting if a door is opened or closed. AC Meter (SCT-013) is a sensor for measuring current intensity. Flex Sensor (custom-made) is a bendable resistor for measuring deflection along its length. Finally, DHT11 is a sensor for measuring temperature and relative air humidity in the environment. With a Wemos D1 mini-embedded system for communicating the sensor input to the central hub for saving data to a local Database.

In Fig. 1, a description of the sensor location in relation to the area where the experiment was carried out is shown. These sensors were placed by the recommendation of a clinical psychology expert to capture relevant data from the subject.

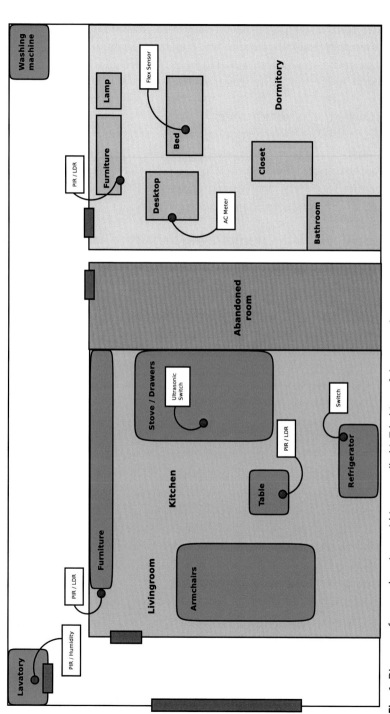

Fig. 1 Diagram of sensor locations within controlled IoT habitat of the test subject.

As an additional measure for the case of failed communication between the sensor node and hub, a buffer was implemented in each sensor node to save data until communication could be established and therefore send all data to the hub to minimize data loss.

2.2.1 Calibration of sensor nodes

As all the sensor nodes were custom-made, a calibration phase was carried out to improve the performance and reduce false negatives. The calibration took approximately 3 weeks. The parameters taken into account were: temperature variations, cloud covering, interior light sources, actions, and duration of activities exerted by the participant.

2.3 Additional data for improving behavioral identification

Along with data captured from the IoT sensor network, two additional data sources relating to the test subject's conduct were also used.

2.3.1 Daily social, personal, and work goals

A mobile phone application was used to register daily a limited number of personal goals (adequate hydration, healthy eating, daily reading, and going for a walk) and social goals (spending time with family, friends, and co-workers, attending meetings, and being more attentive to people), chosen by the individual himself to reflect his lifestyle and daily activities by registering the daily accomplishment, or not, of these 8 general goals (4 personal and 4 social).

2.3.2 Stress, sleep, and heart rate meter

A commercial smartwatch band (Garmin Vivosmart 3®) with the capabilities of a continuous measurement of stress, sleep, and heart rate was used to complement both the IoT sensor network data and daily goals. It must be noted, that although this smartwatch band does not have clinical-grade accuracy, it provides enough consistency to be used for the prototype used in this experiment.

2.4 Data capture format

Three sources of data were established in order to gain sufficient significant data on the study subject's conduct for later processing. The first data from the IoT sensor network, where data captured was saved in a Database under the format shown in Table 2. However in the case of the analog type sensors (Humidity sensor, ultrasonic sensor, LDR sensor, AC Meter sensor), their

Table 2 Data format for captured data from IoT sensor network.

Attribute	Description
Nsensor	Sensor event counter
State	State of sensor, 0 or 1
Time	Timestamp of registered event
idSensor	Sensor identification number

status attribute was set from an activation range which established a final value of 0 or 1 depending on if the activation limit established in each one was exceeded.

The second source of data was from the mobile phone application, where each day a list of chosen goals was registered into the Database. Finally, the third data source was from a commercial smartwatch band, which provided a daily summary value of health-related biometrics attributes: stress, sleep time, and heart rate. Therefore, only one value per day was saved into the Database referencing these attributes.

2.5 Data preparation

The current format, from Table 2, of data could not be directly used in series with clustering algorithms for finding groups of behavioral conduct because some sensors would most likely sense nothing, whereas others would sense much more frequently. Therefore, a data transformation procedure was proposed to improve group sensor data in time frames. These time frames would be within a specific timeframe of 1 h each with a 30-min revision between each one. That is, many, few, or no sensors would usually sense activity, thereby grouping within these timeframes many readings could exist for each data instance. For example, if motion was detected within the kitchen, the refrigerator was opened, the kitchen stove also detected usage, and presence was detected from the kitchen table, all within a timeframe of 1 h, it could be inferred that the participant was active in preparing food and then eating.

During the sensor calibration phase, it was also observed that activities for the participant were restricted to two different time frames, 5:30 to 7:30 h in the morning, and 17:30 to 22:30 h in the evening, then separating the meaningful hours for sensing into only seven 1 h timeframes. Table 3 shows the complete list of time frames used in the dataset. The sensing phase for the

Table 3 Timeframes details.

Attribute name	Description
Date	dd/mm/yyyy for the time frame (only used once per each data instance)
AS	Sum of the time intervals in which an event was detected within 1 h.
FC	Number of intervals in which an event is detected within each set time frame (1 h).
FF	First frame where a valid activity was detected.
LF	Last frame where a valid activity was detected.
TAD	Times an activity was detected.

participant lasted 47 days. In consequence only 329 data tuples would conform the final dataset to be used by pattern mining techniques in order to find some meaningful distinct behavioral conduct groupings.

2.5.1 Attributes for each sensor node

One data instance was constructed by a summary of attributes relating to some relevant information for each sensor. These attributes are named: AS, FC, FF, LF, and TAD. Table 4 shows a brief description of each of these main attributes as well.

A more detailed description for each attribute is given in the following text. In case the same sensor performs a detection, for example: of movement, several times within the same time frame, AS value will be increased based on a sum between the intervals in which the time frame was divided, in this case 1 h was divided into intervals of 30 min obtaining two frames to add each other, if the sensor detected an event in the first lapse the final value will be 1, if it was detected in the second the value would be 2 and finally if it was detected in both lapses the AS value would be 3.

As a validation attribute, an accumulator was added to control the number of time frames that the sensor node detected valid activity, if the number of frames provided a value greater than one, it implied that in more than one time frame an event had been detected. Unlike the AS attribute in which the interval in which an event was detected can be determined, the current attribute provides the number of frames involved, with the intention of complementing the result obtained by AS and facilitating the monitoring of the chronology developed by the individual, this attribute was given the name of FC, for Frame Counter.

Table 4 Summary of attribute descriptions after data processing, for each sensor node.

Time of day	Time interval (h)
Morning	5:30 to 6:29
	6:30 to 7:30
Midday	Participant is at work, no sensing is carried out
Evening	17:30 to 18:29
	18:30 to 19:29
	19:30 to 20:29
	20:30 to 21:29
	21:30 to 22:30

Another validation attribute is for the first sub-timeframe for which a valid activity was recorder, that is, if within a timeframe (of 1 h), there exists a total of 2 sub-timeframes, each one 30 min long, when an activity was first detected, e.g., on 6:43 h then this first frame would be equivalent to the first sub-timeframe, consequently this attribute would have a value of 1. This attribute is called FF, for First Frame.

Analogous to the attribute FF, another value is extracted which represents the last sub-timeframe within a timeframe, e.g., on 7:27 h was the last time an activity was detected by any of the sensors, then the last sub-timeframe would have a value of 2. This attribute is called LF, for Last Frame.

The last attribute is a value which counts the amount of times a given sensor detected an activity within a timeframe, then divided by 2. This division by 2 is to compensate duplicate detections by the sensor, i.e., any given sensor registers a change in state, for when it detected an activity, and for when it stopped detecting, therefore the division by 2 is to register one detection for the complete detection cycle (change from no detection to detection, then back to no detection). This attribute is called TAD, for Times Activity Detected.

2.5.2 Final preprocessing

The last data manipulation step before applying pattern mining is to build a complete and integrated data row for all sensors. All 13 sensors had to be identified and differentiated as well, Table 5 shows a summary of naming conventions used for each one.

Table 5 Naming conventions for sensors.

Short name	Long name	ID
LE	Light for main entrance	0
M	Motion in living room	1
R	Refrigerator door	2
E	Near cooking stove	3
G	Pantry drawer	4
UT	Motion over kitchen table	5
UL	Ceiling light in kitchen	6
B	Motion in lavatory	7
BH	Temperature and humidity in lavatory	8
LC	Light in bedroom entrance	9
MC	Motion in bedroom	10
CC	Presence in bed	11
EC	Use on computer	12

Afterward, a data tuple must be constructed for each attribute group (AS, FC, FF, LF and TAD), for each sensor, that is, for all 13 sensors, considering 5 attributes for each one, a total of 65 attributes for each timeframe.

2.6 Attribute reduction

Considering the high number of attributes created up to now, it would be beneficial to reduce the number as having such a high number of attributes could add unnecessary complexity to the pattern mining process. Therefore a Principal Component Analysis (PCA) was used on the dataset to reduce the number of attributes, from 65 down to 10. However, a detailed analysis of the selected attributes (M_FC, BH_FF, G_FC, G_AS, CC_FF, UT_AS, EC_AS, E_LF, R_FF, MC_TAD) revealed the need to eliminate those that represented events that the individual performed in a limited number of times, e.g., all attributes related to EC, registering multiple inactivity values for the rest of the time and therefore affecting the selection of attributes by the PCA algorithm. Therefore, for this work the reduction of attributes was performed from the elements selected by the PCA algorithm and its subsequent individual analysis for the purpose of evaluating its content, in case it presented mostly inactivity values (UL, EC, G), the column would be discarded. Once the tuples with a higher number of inactivity were eliminated, a correlation matrix was used with the intention of analyzing the elements of each attribute (AS, FC, FF, LF and TAD) with the highest correlation among the rest. Finally, our dataset consisted of (UT_AS, M_FC), obtained from the first use of the PCA algorithm, (LE_AS, R_AS, R_TAD, E_AS, BH_AS, LC_AS, MC_CF, CC_AS), obtained by using the correlation matrix in conjunction with the PCA algorithm, (Social, Goals, Stress, Heart_rate, Sleep) were attributes incorporated by the use of the mobile devices by the individual (smartwatch and mobile app) and finally an extra attribute was added (Time) with the intention of facilitating the visualization of the data points grouped by the final model, however this last one was not used in the training dataset.

2.7 Final data cleansing

Data tuples with mostly zeros were discarded as they did not represent a sufficient amount of activity within those timeframes, thus reducing the final tuple count from 329 to 197. Data from the two other sources of information was also used in conjunction with the final dataset obtained after the sensing phase, that is, for each of the seven timeframes per day, a mean value corresponding to the attributes stress, sleep and heart rate was added and on the other hand a weighted value obtained from the attributes goals and social was assigned starting from the fifth timeframe of the day. Finally, one last attribute was added, and it is a value from 1 to 7, which represents the timeframe for which the tuple represents, as shown in Table 3, in that order, all data was normalized to a unit norm.

3. Behavioral anomaly pattern mining results

To mine human behavioral anomalies based on the previously constructed dataset, first a correlation matrix was built based on all attributes in order to briefly analyze the relationship between attributes, as shown in Fig. 2, where higher values represent increased correlation and a lower/negative values represent a highly negative correlation.

This correlation matrix shows a positive relationship between activities performed in the area of the kitchen, as well between the lavatory and the bedroom of the participant. Also, the attributes for *Goals* and *Social* also showed great correlation between them, in addition to *Stress* and *Heart rate*.

	LE_AS	M_CF	R_AS	R_TAD	E_AS	UT_AS	BH_AS	LC_AS	MC_CF	CC_AS	Social	Goals	Stress	Heart_rate	Sleep	Time
LE_AS	1	0.82	0.43	0.17	0.24	0.16	-0.042	0.21	0.07	-0.44	-0.03	-0.032	0.061	0.049	-0.018	-0.032
M_CF	0.82	1	0.55	0.35	0.39	0.31	0.063	0.23	0.1	-0.47	0.074	0.07	0.022	0.053	-0.038	-0.25
R_AS	0.43	0.55	1	0.8	0.56	0.54	-0.098	0.15	-0.039	-0.44	-0.09	-0.11	0.095	0.21	0.07	-0.25
R_TAD	0.17	0.35	0.8	1	0.5	0.61	0.017	0.057	-0.1	-0.35	0.02	-0.021	-0.028	0.11	0.075	-0.096
E_AS	0.24	0.39	0.56	0.5	1	0.32	-0.065	0.085	-0.0026	-0.32	-0.061	-0.086	0.016	0.077	0.043	-0.24
UT_AS	0.16	0.31	0.54	0.61	0.32	1	0.024	0.0019	-0.11	-0.22	0.02	0.00026	0.13	0.21	0.11	0.039
BH_AS	-0.042	0.063	-0.098	0.017	-0.065	0.024	1	0.14	0.34	0.073	1	0.96	0.012	0.057	0.0092	0.25
LC_AS	0.21	0.23	0.15	0.057	0.085	0.0019	0.14	1	0.56	-0.22	0.16	0.16	0.11	0.071	-0.10073	-0.046
MC_CF	0.07	0.1	-0.039	-0.1	-0.0026	-0.11	0.34	0.56	1	-0.24	0.34	0.34	0.035	0.033	0.019	0.1
CC_AS	-0.44	-0.47	-0.44	-0.35	-0.32	-0.22	-0.073	-0.22	-0.24	1	-0.079	-0.032	0.07	0.0088	0.0067	0.18
Social	-0.03	0.074	-0.09	0.02	-0.061	0.02	1	0.16	0.34	-0.079	1	0.97	0.0031	0.06	0.0072	0.24
Goals	-0.032	0.07	-0.11	-0.021	-0.086	0.00026	0.96	0.16	0.34	-0.032	0.97	1	-0.027	0.048	-0.0097	0.25
Stress	0.061	0.022	0.095	-0.028	0.016	0.13	0.012	0.11	0.035	0.07	0.0031	-0.027	1	0.66	-0.025	0.0097
Heart_rate	0.049	0.053	0.21	0.11	0.077	0.21	0.057	0.071	0.033	0.0088	0.06	0.048	0.66	1	0.11	0.032
Sleep	-0.018	-0.038	0.07	0.075	0.043	0.11	0.0092	-0.0073	0.019	0.0067	0.0072	-0.0097	-0.025	0.11	1	0.039
Time	-0.32	-0.25	-0.25	-0.096	-0.24	0.039	0.25	-0.046	0.1	0.18	0.24	0.25	0.0097	0.032	0.039	1

Fig. 2 Correlation matrix between attributes.

Lastly, there is also a strong correlation between the attributes of *Social* and *Goals* to *BH_AS* (Anomaly Scale in Lavatory), which is the one that corresponds to taking a shower.

3.1 Behavioral patterns through data clustering

A K-Means algorithm was used in order to find the behavioral patterns within the dataset which has been prepared up to now. In order to find the optimal value of behavioral pattern groups, the Elbow Method was used, as shown in Fig. 3, where the optimal value is 2 groups.

Various indices were also used in order to best assess the performance, among which were used are: Silhouette, Calinski-Harabasz, Davies-Bouldin, and Dunn. For which results are shown in Table 6, where the best values were obtained with $K = 2$. Therefore 2 groups are considered to be optimal.

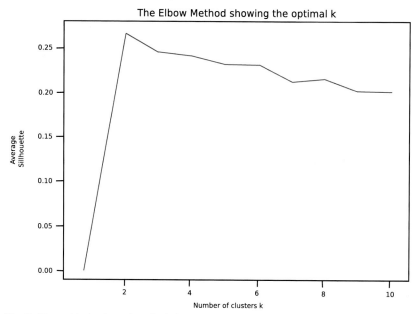

Fig. 3 Elbow Method used to find the optimal K value for K-Means algorithm.

Table 6 Performance indices obtained with K-Means on constructed dataset.

Index	2	3	4	5	6	7
Silhouette	0.278130	0.245662	0.240829	0.231073	0.235537	0.229549
Calinski-Harabasz	73.135557	63.318323	59.460458	53.719192	50.208738	46.467218
Davies-Bouldin	1.539407	1.562347	1.565273	1.381216	1.488740	1.457854
Dunn	1.81431	0.203615	0.123841	0.140905	0.154066	0.140905

Although by identifying two groups, by no means can one know what they represent, just as it is not possible to identify the apparent anomalies, this still requires further analysis, which will now be detailed.

3.1.1 Per group analysis
A quick analysis was made to assess if groups were chosen mainly on morning and evening activities, for this, a simple visualization of what timeframes were part of each group. It was found out that both groups included data

from all seven timeframes, discarding the idea that one group would be comprised of morning activities and the other group of evening activities. This was not the case.

It was also found out that some activities were skewed from one group to the other, that is, in the first group most activities were carried out inside the bedroom and the lavatory, this being interpreted as the participant being asleep or getting ready to go to work in the morning hours. In retrospect, the other group held activities mainly carried out in the living room, kitchen as well as the lavatory. These activities were mainly for when the participant was eating breakfast, arrived from work and promptly prepared dinner and ate. Although it must be noted that for both groups, these activities did conform to the majority of them, but not exclusively related to them. Fig. 4 shows the general tendency of timeframe distribution per each found group. It can be seen that both groups are conformed from tuples from all timeframes, except for the case of cluster 1, where no tuples exist from the timeframe of 5:30 to 6:29.

Another finding is that the first group, for attributes pertaining to _AS, values were higher, that is, the first groups had more tuples where the participant was more active during the day. In contrast to the second group which had a lower value within these attributes, meaning that this second cluster has more tuples with timeframes with less activity.

3.2 Behavioral anomaly pattern mining

As a K-Means algorithm provided two clusters which conform two behavioral groups, the centroids for these two clusters were also useful in finding outliers in the data, in other words, outliers in the data would represent behavioral anomaly patterns within the two found groups. This was achieved by obtaining descriptive statistical values such as average, median, mode, minimum and maximum, from data belonging to each cluster. These values are shown in Table 7.

A histogram based on the distances measured for each datum for both clusters, is shown in Fig. 5. Where an overlap for both groups can be visualized, and provides a quick view into possible outliers.

To find the outliers, a box plot technique was used by using the same datum distances. Fig. 6 shows the boxplot results along with its descriptive data, it can also be noted that the first cluster group has a higher skewness in comparison with the second cluster group, and that the first has more extreme apparent outliers while the second does not, although both do have outliers.

Cluster elements 0		Cluster elements 1	
Hours	Number of elementes in cluster	Hours	Number of elementes in cluster
5:30-6:29 |	17	5:30-6:29 |	0
6:30-7:30 |	12	6:30-7:30 |	23
17:30-18:29 |	11	17:30-18:29 |	11
18:30-19:29 |	6	18:30-19:29 |	21
19:30-20:29 |	24	19:30-20:29 |	8
20:30-21:29 |	28	20:30-21:29 |	5
21:30-22:30 |	30	21:30-22:30 |	1

Fig. 4 Distribution of time frames per each found cluster group.

Table 7 Statistical analysis per cluster.

Cluster	Average	Median	Mode	Min	Max
Cluster 0	0.34497	0.30745	0.25105	0.12140	0.85545
Cluster 1	0.41464	0.43104	0.51365	0.20785	0.62794

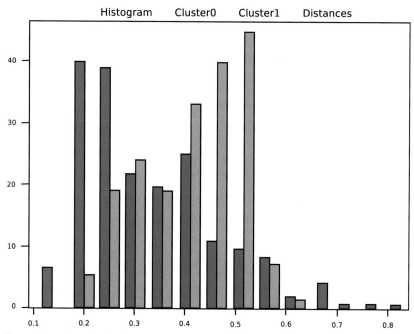

Fig. 5 Histogram for both cluster groups based on datum distances from centroids.

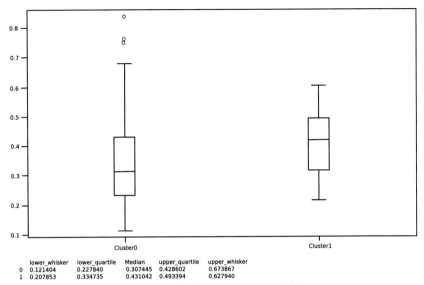

	lower_whisker	lower_quartile	Median	upper_quartile	upper_whisker
0	0.121404	0.227840	0.307445	0.428602	0.673867
1	0.207853	0.334735	0.431042	0.493394	0.627940

Fig. 6 Boxplot graph for both found clusters and detailed information.

	LE_AS	M_CF	R_AS	R_TAD	E_AS	UT_AS	BH_AS	LC_AS	MC_CF	CC_AS	Social	Goals	Estress	Heart_rate	Sleep
69	0	2	3	8	3	3	0	0	0	0	1	1	0.9	5.2	7.9
110	0	0	3	8	2	0	0	1	1	0	1	1	2.0	5.4	8.6
114	0	0	0	0	0	0	0	1	1	1	1	1	2.0	5.4	8.6
122	0	2	3	5	3	0	0	0	0	0	1	1	1.7	5.1	4.7
182	0	1	3	12	0	3	2	2	1	0	3	2	0.6	5.1	8.2
59	2	1	2	2	2	0	0	2	1	0	1	1	3.8	5.6	7.6
91	0	0	0	0	0	0	0	0	0	3	1	1	2.6	5.2	3.8

Fig. 7 Found outliers and their representative attribute data.

Only 7 outliers were found using the boxplot technique to find them. Fig. 7 shows a summary of distances and data tuples information which belong to these found outliers.

Finally, to provide a more robust answer for finding behavioral anomaly patterns using the concept of outlier detection, previous results from the boxplot technique were contrasted with another outlier detection technique named Copula-Based Outlier Detection (COPOD). This technique, in contrast to the boxplot approach, found 40 outliers, but only the common outliers between both techniques were of interest, that is, between the 40 outliers found by COPOD and the 7 found by using the boxplot approach, only 5 were found to be in common. These common outliers were deemed to be the best representation and are shown in Fig. 8.

	LE_AS	M_CF	R_AS	R_TAD	E_AS	UT_AS	BH_AS	LC_AS	MC_CF	CC_AS	Social	Goals	Estress	Heart_rate	Sleep	Date	Time	cluster_
69	0	2	3	8	3	3	0	0	0	0	1	1	0.9	5.2	7.9	9	4	1
91	0	0	0	0	0	0	0	0	0	0	1	1	2.6	5.2	3.8	18	1	0
110	0	0	3	8	2	0	0	1	1	0	1	1	2.0	5.4	8.6	21	3	1
122	0	2	3	5	3	0	0	0	0	0	1	1	1.7	5.1	4.7	23	4	1
182	0	1	3	12	0	3	2	2	1	0	3	2	0.6	5.1	8.2	7	5	1

Fig. 8 Common outliers between boxplot approach and COPOD.

These outliers can be said to be true behavioral anomaly patterns as found through the process of pattern mining of an IoT environment.

4. Discussion

After analyzing the final obtained group of outliers, shown in Fig. 8, the following statements can be interpreted in relation to human anomalous behaviors detected by the participant's activities within his home environment:

The five found outliers represent remarkable events that together merit its selection for having unique events in comparison with everyday activities, that is, events such as opening a higher number of times the refrigerator; moving within the home during non-normal hours (Hours out of a range established according to the activity performed by the specific individual); presenting stress levels outside the median of (2.25), with Q1 (1.5), Q2 (3.65) and a full range of [0.65–6.5] for stress levels; carrying out activities which were outside the participant's daily routine; or even presenting alterations in his sleep cycle (e.g., 3.8 h) where Q2 was 7.65 h, Q1 (7.1 h), Q3 (8.2 h) and a full range of [5.7–9.4] h. These events and different ranges were specific to the participant, assessed and calculated through the proposed methodology.

An essential part of this proposal is the presentation of results by the final pattern mining process since it not only shows the single value that stands out from the rest, i.e., individual outliers, but also presents the complete chronology in which this data is involved, providing insight on the context in which the individual carries out his day-to-day activities. This, is through the identification of attribute significance and meaning.

Reducing the attribute count from 65 to 10 is very significant, since future systems of similar nature can be optimized to use only the minimum required sensors, reducing the cost and complexity of such systems. Although for each new individual where a similar natural system would be used, an initial trial and data collection phase would be first required to suit the individual, where the proposed methodology utilizing pattern mining and outlier detection would decide, in the end, all meaningful attributes. This system is not a fixed solution, but through the proposed methodology, each system can be custom-made to suit an individual.

The presented work centers on providing a working methodology for human behavioral anomaly pattern mining within an IoT environment, for which, after reviewing final results, we can confirm with certain assurance that obtained results offer sufficient evidence to conclude that human behavioral anomalies were in fact detected through the proposed methodology, therefore setting precedence that pattern mining can be used for this purpose.

By automating the pattern mining methodology from data source acquisition to outlier detection, a model of the individual could potentially detect real-time behavioral anomalies, whereas it could be potentially applied under contexts such as suicide prevention or the discovery of other undiagnosed mental disorders that an individual may present throughout his/her life. Moreover, by continually monitoring data for the individual and periodically updating the outlier detection model, this system could adapt through time to the individual.

5. Conclusion

This study aimed to explore the viability of detecting anomalous behaviors in one individual in a case study based solely on his interactions with sensors in his home environment as well as an app that provided self-reported data and a smartwatch band.

The main results of this experiment indicate that it is possible to transform obtained data between the relationship of an individual and their environment using IoT technology in order to create a pattern of behaviors in relation to time that allows through the use of pattern mining to detect the anomalous behaviors within these as well as the information that relates to their selection by the model, this final dataset, promises to complement the information that could support an expert in the area of mental health to provide a more objective psychological follow-up according to the individual who requests it.

An IoT environment was designed with sensors selected for significant activity monitoring related to behavioral patterns based on general psychological analysis of relevant behaviors. With the sensor network being designed in-house, data capture was direct, and we had more control over any quick changes that arose during the calibration phase.

A data arrangement process is proposed that tailors the monitored data from the IoT environment and preprocesses it into a simplified dataset for simpler pattern recognition processing, thus facilitating the pattern mining for detecting outliers. Even though many trials were carried out in order to find the best preprocessing, the one shown in this paper provided the best results.

The exploratory objective of this study was achieved, such that the detection of anomalous behaviors in individuals based solely on their interactions with sensors in their home environment and other monitoring technologies was successfully carried out. These results open the door to continue exploring alternate technologies, sensors, and ML techniques, and use cases with multiple participants within the same home environment.

References

[1] American Psychiatric Association, Diagnostic and Statistical Manual of Mental Disorders, Author, Arlington, VA, 2013.

[2] D.J. Hughes, Psychometric validity: establishing the accuracy and appropriateness of psychometric measures, in: P. Irwing, T Booth, D.J. Hughes (Eds.), The Wiley Handbook of Psychometric Testing: A Multidisciplinary Approach to Survey, Scale and Test Development, John Wiley & Sons Ltd., Chichester, UK, 2018, pp. 751–779. https://onlinelibrary.wiley.com/doi/10.1002/9781118489772.ch24.

[3] N. Martinez-Martin, T.R. Insel, P. Dagum, H.T. Greely, M.K. Cho, Data mining for health: staking out the ethical territory of digital phenotyping, NPJ Digit. Med. 1 (1) (2018) 1–5.

[4] T. Yarkoni, J. Westfall, Choosing prediction over explanation in psychology: lessons from machine learning, Perspect. Psychol. Sci. 12 (6) (2017) 1100–1122, https://doi.org/10.1177/1745691617693393.

[5] G. Azar, C. Gloster, N. El-Bathy, S. Yu, R.H. Neela, I. Alothman, Intelligent data mining and machine learning for mental health diagnosis using genetic algorithm, IEEE Int. Conf. Electro. Inf. Technol. 2015 (2015) 201–206.

[6] R.A. Bernert, A.M. Hilberg, R. Melia, J.P. Kim, N.H. Shah, F. Abnousi, Artificial Intelligence and Suicide Prevention: A Systematic Review of Machine Learning Investigations, Int. J. Environ. Res. Public Health 17 (16) (2020) 5929.

[7] H. Stringer, Tapping into a New Tool for Research, American Psychological Association, Washington, D.C., 2021.

[8] C. Stachl, et al., Predicting personality from patterns of behavior collected with smartphones, Proc. Natl. Acad. Sci. U. S. A. 117 (30) (2020) 17680–17687.

[9] R.R. McCrae, P.T. Costa, Validation of the five-factor model of personality across instruments and observers, J. Pers. Soc. Psychol. 52 (1) (1987) 81–90.

[10] W. Bleidorn, C.J. Hopwood, Using machine learning to advance personality assessment and theory, Pers. Soc. Psychol. Rev. 23 (2) (2019) 190–203.

[11] J. Golbeck, C. Robles, K. Turner, Predicting personality with social media, in: Conference on Human Factors in Computing Systems - Proceedings, 2011, pp. 253–262.

[12] H.K.M. Meeren, C.C.R.J. Van Heijnsbergen, B. De Gelder, Rapid perceptual integration of facial expression and emotional body language, Proc. Natl. Acad. Sci. U. S. A. 102 (45) (2005) 16518–16523.

[13] I. Behoora, C.S. Tucker, Machine learning classification of design team members' body language patterns for real time emotional state detection, Des. Stud. 39 (2015) 100–127.

[14] N. Kojovic, S. Natraj, S.P. Mohanty, T. Maillart, M. Schaer, Using 2D video-based pose estimation for automated prediction of autism spectrum disorders in young children, Sci. Rep. 11 (1) (2021) 1–10.

[15] R.K. MacKinnon, Seeing the Forest for the Trees: Tree-Based Uncertain Frequent, Pattern Mining, Springer International Publishing, 2014. https://mspace. lib.umanitoba.ca/items/e253d863-ab99-4412-bdac-3722eca43aa7.

[16] G. Orrù, M. Monaro, C. Conversano, A. Gemignani, G. Sartori, Machine learning in psychometrics and psychological research, Front. Psychol. 10 (2020) 2970.

[17] P. Lea, Internet of Things for Architects : Architecting IoT Solutions by Implementing Sensors, Communication Infrastructure, Edge Computing, Analytics, and Security, Packt Publishing Ltd, 2018.

[18] A. Cesta, G. Cortellessa, M.V. Giuliani, F. Pecora, M. Scopelliti, L. Tiberio, Psychological implications of domestic assistive Technology for the Elderly, PsychNology J. 5 (3) (2007) 229–252.

[19] H. Sundmaeker, P. Guillemin, P. Friess, S. Woelfflé, Vision and challenges for realising the internet of things, Clust. Eur. Res. Proj. Internet Things, Eur. Commision 3 (3) (2010) 34–36.

[20] A. Souri, M.Y. Ghafour, A.M. Ahmed, F. Safara, A. Yamini, M. Hoseyninezhad, A new machine learning-based healthcare monitoring model for student's condition diagnosis in internet of things environment, Soft Comput. 24 (22) (2020) 17111–17121.

About the authors

Rosario Sánchez-García is a software developer in the medical field and holds a master's degree from the Universidad Autónoma de Baja California (UABC) in engineering with a specialization in machine learning and embedded systems for monitoring livable environments to detect behavior patterns.

Alejandro Dominguez-Rodriguez is a graduate in Psychology from the Autonomous University of Ciudad Juárez. Received the Santiago Grisolía's scholarship from the Valencian Community, Spain and he did his master's and doctorate degrees at the University of Valencia. He works as an assistant professor at the Department of Psychology, Health, and Technology at the University of Twente, the Netherlands. He conducts research in the implementation of user-centered interventions supported by technologies such as Virtual Reality, Augmented Reality, Virtual Assistants, Machine Learning, Serious Games, self-applied treatments through web platforms, and pedometers, among others. He has the distinction of the National System of Researchers, Level I, and is Academic Editor in the journal PLOS ONE and Review Editor in Frontiers in Digital Health and Ethical Digital Health. He directs the Internet Treatments for Latin America and Spain (ITLAS) group www.itlasgroup.com.

Violeta Ocegueda-Miramontes is currently working as a full-time professor and researcher in the Faculty of Chemical Sciences and Engineering at Autonomous University of Baja California (UABC). She holds a PhD from the same University in the area of Computer Sciences. She has published in the subfield of Knowledge Management. Her current research interests are in the areas of artificial intelligence.

Leocundo Aguilar received the PhD degree in computer science from the Universidad Autónoma de Baja California (UABC). He is currently a full-time professor with the Computer Engineering program, UABC. His current research interests include embedded systems, wireless sensor networks, and intelligent systems applied to ubiquitous computing. He is also a member of the National System of Researchers (SNII).

Antonio Rodríguez-Díaz holds a PhD and is currently a Faculty Member at Universidad Autónoma de Baja California since 1993.

Sergio Cervera-Torres is a psychologist and researcher currently affiliated to the Leibniz-Institut für Wissensmedien (IWM) in Tübingen, Germany. He is also an associated lecturer at the Open University of Catalonia (UOC), Barcelona, Spain. His research interests fold into the area of the Positive Technology field and associated psychological constructs (e.g., resilience). Concretely, he has focused on emotion and cognitive processes linked to Human-Computer Interaction (HCI) and eHealth. He has been awarded with the prestigious Deutsche Forschungsgemeinschaft (DFG, German Research Foundation) research grant.

Mauricio A. Sanchez holds a PhD from Universidad Autonoma de Baja California and carried out a postdoctoral fellowship at Tecnologico Nacional de Mexico Tijuana. He specializes in machine learning and applied artificial intelligence. Has published in various high impact journals, as well as multiple books. He has also guided various postgraduate thesis in the areas of applied artificial intelligence. He currently is a professor and researcher at Universidad Autonoma de Baja California and belongs to the National System of Researchers (SNII).

CHAPTER THREE

Indoor localization technologies for activity-assisted living: Opportunities, challenges, and future directions

Muhammad Zakir Khan, Muhammad Farooq, Ahmad Taha, Adnan Qayyum, Fehaid Alqahtani, Adnan Nadeem Al Hassan, Kamran Arshad, Khaled Assaleh, Shuja Ansari, Muhammad Usman, Muhammad Ali Imran, and Qammer H. Abbasi
James Watt School of Engineering, University of Glasgow, Glasgow, United Kingdom

Contents

Advances in Computers, Volume 133
ISSN 0065-2458
https://doi.org/10.1016/bs.adcom.2023.11.001

Abstract

Remote health monitoring is becoming more and more desirable as the healthcare industry transitions from hospital-focused to home-focused care, especially with an aging population. The reliability of wireless sensors and communication systems are gaining more trust on monitoring patients' daily activities in their homes. However, for such systems to be effective, localizing the activities performed in an indoor environment is essential, especially for making decisions during critical moments such as walking or falling, which can lead to mortality and immobility. This chapter explores home-centric activity localization systems (ALSs), covering enabling technologies, challenges, and possible future directions. Additionally, the chapter provides an extensive overview of indoor localization systems that rely on channel state information and analyzes the approaches, techniques, and technologies used for indoor localization. Finally, the chapter compares the results and performance of existing ALSs, followed by a discussion on the technological implementation, open challenges, possible solutions, and future directions.

1. Introduction

Indoor localization has recently grown in popularity as a result of its extensive usage in numerous applications, including battlefield surveillance, intelligent traffic, disaster prediction, and indoor navigation [1–3]. These systems have shown to be very useful in providing effective indoor localization systems (ILSs) and home care for the elderly. The elderly population is growing along with life expectancy due to advancements in disease diagnosis and treatment, while hospital capacity is rapidly decreasing [4]. According to the United Nations (UN), there will be 2.1 billion more elderly people on the earth by 2050 [5]. This exemplifies the significance of technology utilization in elder care for those who are losing their autonomy but still wish to stay in their own homes and need continuing, real-time assistance. Therefore, it is critical to pay attention to the elder patient's daily activities. The main focus is the elderly patient, whether they are at home or in a nursing facility. The system must be able to detect anomalies from normal behavior as well as abnormalities that indicate a decline in patient's abilities. Accurate and trustworthy indoor localization and tracking systems are becoming both necessary and desired in healthcare monitoring to ensure the efficacy of these systems. Intelligent home care systems that significantly reduce hospitalization and save healthcare costs are now possible due to advancements in artificial intelligence, wireless communication, and nanotechnology.

It is imperative for healthcare facilities to provide specialized care to vulnerable patients such as infants, elderly patients, and those with physical or mental disabilities. To ensure that patients receive the necessary care and assistance, medical professionals must have knowledge of their indoor location [6]. Wireless sensing technology, which includes transmitter and receiver nodes, is a viable method for locating patients within buildings [7,8]. During the last decade, a wide range of assistive technologies have emerged in response to the ambient intelligence paradigm. According to this paradigm, around 90% of elderly patients prefer to receive nursing care at home [9]. The activity-assisted living (AAL) systems aim to provide the elderly with a smart environment that offers proactive assistance through natural interaction. These systems employ various monitoring devices that can react and interact with users in response to their situation. One of the critical components of AAL systems is activity recognition such as smart homes, video monitoring systems, and caregiving robotics. Despite the fact that significant research in this area is still difficult to find a balanced solution that satisfies the demanding requirements of an AAL system, such as being built on low-cost equipment, providing highly accurate information, and having increased user acceptance while ensuring compliance to end-user privacy rules.

Several indoor localization technologies (such as inertial positioning [10], Bluetooth [11], Light [12], ultrasound [13], and visible and Wi-Fi [14]) have recently made significant advancements and can be used alone or in combination (using a technique known as sensor fusion [15]) for AAL. Research focused on these technologies for ILS including WiFi, radio frequency identification (RFID), Bluetooth, ZigBee, ultrawide band, and vision. While comparing and evaluating ILS technology, the primary factors to consider are range, power requirements, cost, latency, and accuracy. Depending on the technology used, a range of signal metrics are used to pinpoint the target's location, including the received signal strength indicator (RSSI), the CSI, fingerprinting analysis, angle of arrival (AoA), time of flight (ToF), time difference of arrival (TDoA), return ToF, and phase of arrival. For instance, researchers in Ref. [7] analyzed WiFi, Bluetooth, and ZigBee as ILS technologies for healthcare sector and evaluated RSSI, ToA, and fingerprint analysis as localization methods. According to their study, the optimal ILS techniques in a healthcare environment are WiFi and fingerprint analysis. Since WiFi-based systems are among the most popular infrastructure due to either being widely available or being capable of being installed quickly and affordably. Furthermore, owing to the simplicity of hardware deployment, WiFi-based indoor localization is a popular choice for offering accurate and efficient location-based services. However, approaches based on fingerprint and geometry are used in the majority of positioning categories. The geometry-based positioning methodology is composed of two distinct approaches: angle-based and range-based. The previous approach is dependent on signals from base stations having angle estimation capability [16]. In order to estimate a position, a significant number of base stations are used to calculate the distance between each target and base station [17]. However the fingerprint-based locating system's components are both offline and online. The offline component's key features are data collection, preprocessing, and the creation of a fingerprint library, while the online component compares real-time signals using algorithms, to identify the mobile terminal. Fig. 1 illustrates how the indoor localization system for AAL can be classified depending on size, type, environment, techniques, approach, and technology.

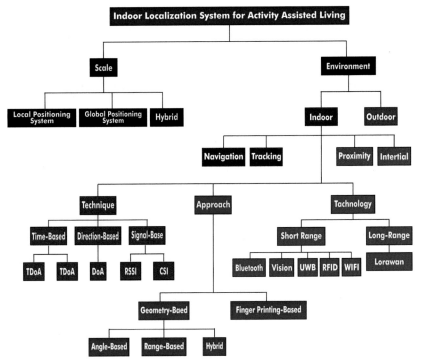

Fig. 1 Indoor localization technology and algorithm classification for activity-assisted living.

2. Indoor localization approaches and techniques

This section covers several approaches, techniques, and strategies for locating objects in an indoor environment. We will be discussing these methods by categorizing them into two categories, i.e., geometry-based and fingerprint-based approaches, which are described next.

2.1 Geometry-based indoor localization approaches

Geometry-based positioning approach has further two categories, i.e., angle-based and range-based methods, which are described below.

2.1.1 Angle-based approach

The angle-based approach determines the user's location by measuring the angle at which the signal from known transmitters is received. If the user device and beacon stations use directional antenna technology, the angle of the transmission can be precisely determined. It determines the user's location by using geometrical information. The user's location may be determined using either the angle of the received signal or the distance to the established known measurement locations.

The directions for locating a person using distance and angle information are shown in Fig. 2. Assuming that there are three fixed wireless stations with known locations where A, B, and C are the nodes. If the user point's distance from each of these stations is known, it is possible to locate them by examining at the intersection of the three circles, as shown in Fig. 2A. Similarly, if

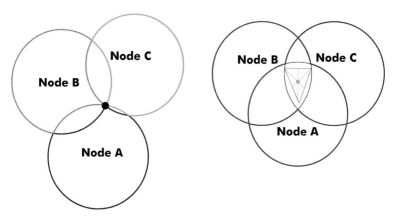

Fig. 2 Determination of the unknown location of three known positions at the intersection of node A, node B, and node C.

we know the angle between the user point and the base stations or the angle between the user point and the three vectors, we can easily locate the user point as shown in Fig. 2B. To ensure the accuracy and reliability of the system, it is recommended that at least three anchor nodes be placed around the object; as the number of anchor nodes increases, the accuracy of the system also increases [18]. It is desirable to use array signal-based angle measuring technology for CSI positioning based on the angle. The CSI measurements made on a WiFi device in accordance with the array antenna angle measurement technique can be used to get phase information that can be used to estimate the AoA, while knowing the positions of two or three RPs is sufficient to locate a moving object in both 2D and 3D space. Using this approach, the distance between the RPs and the angles may be used to determine the transmitter's location. Although the issue of multipath is mitigated by adding multiple antennas, there is a significant correlation between the phase differences of the signals received by each antenna on the access point in CSI-based AoA estimation. The accuracy of geo-location estimate results depends on the CSI phase information used as modern commercial Wi-Fi networks often do not have additional antennas,

Schmidt [19] developed that the MUSIC algorithm employs a network of several antennas to achieve high levels of resolution, estimate accuracy, and stability. Nevertheless, commercial WiFi equipment often only has a few antennae, making it challenging to overcome indoor multipath. Xiong et al. [20] developed the ArrayTrack hardware platform for indoor locating utilizing CSI using six to eight antennas. This platform's positioning accuracy range is between 30 and 50 cm. In order to calculate AoA with high accuracy, it makes use of spatial smoothing, array geometric weighting, and AoA spectrum synthesis methods. The Phaser system, suggested by Ref. [21], measures angles using CSI amplitude and phase data. The system eliminates the phase difference between networking cards and shares one antenna among the remaining five antennas in order to achieve an average positioning error of less than 1 m. Zhang et al. [22] proposed the 3D-WiFi, which employs an L-shaped array to measure angles and is a three-dimensional indoor positioning system. Comparing the 3D-WiFi sparse reconstruction method to the MUSIC, it is claimed to provide decimeter-level precision three-dimensional position accuracy in open indoor environment.

2.1.2 Range-based approach

The accuracy of the RSSI-based positioning approach is significantly degraded by strong multipath effects and indoor shadows. Moreover, positioning accuracy is further effected by the fluctuation in RSSI measurements caused by a number of factors including background noise and human movement. To overcome these issues, researchers are focusing on using CSI to improve the poor spatial stability and low accuracy associated with RSSI. The *FILA* system was established in 2012 by Wu et al. [23] to make a relationship between CSI and the distance. This system converts CSI amplitude into a valid CSI value using a complicated indoor propagation model and a fast training method based on supervised learning. The median accuracy achieved by *FILA* in the research lab, university classroom, and corridor environments is 0.45 m, 1.2 m, and 1.2 m, respectively. In 2018, a new location tracking system called *WiCapture* was proposed by Kotaru and Katti [24] that incorporates a wireless channel model that accounts for frequency offset, phase distortion, and channel change to determine the transmitter's displacement between subsequent CSI by computing the AoD and loss of each path. It achieves a median trajectory error when used within offices of 0.88 cm, 1.51 cm, and 0.85 cm, respectively. Similarly, the *LiFS* system developed by Ref. [25], a device-free localization system based on accurate models and requiring minimum human effort. Depending on whether the target is in the First Fresnel Zone, *LiFS* uses a number of preprocessing methods to the raw CSI in order to construct a power fading model equation. LiFS can accurately predict the destination location in both line-of-sight (LoS) and non–line-of-sight (NLoS) scenarios by resolving a set of PFM equations created for each path, with median accuracy values of 0.5 m and 1.1 m, respectively. Moreover, Han et al. [26] presented the *S-Phaser* as a single Wi-Fi AP-based ILS that is able to localize the user with an accuracy of 1.5 m by first eliminating phase and angle errors with the interpolation elimination technique, calculating the transmission distance with broadband angle ranging, and then using triangulation to locate the user in the LoS scenario.

2.1.3 Hybridization of angle and range-based approach

Researchers are focusing on the integration of angle and range measurement, as relying solely on angle or range measurement for positioning has its advantages and disadvantages.

In 2017, Han et al. [27] introduced a new indoor localization approach that uses a single access point to pinpoint the exact position of the user's terminal. They acquire CSI from several channels and remove the phase error in the CSI using the interpolation elimination technique. The CSI has been processed using the spatial smoothing method and the expected values of

AoA and ToA. The LoS clustering method is then used to determine the direct path. This method produces a median inaccuracy of 0.6 m in static environments with straight paths. Similarly, Ahmad et al. [28] proposed a modified matrix pencil (MMP) approach to estimate the AoA and ToF of all dominant multipath components using WiFi CSI data. Compared to a two-dimensional multiple signal classification method with a similar estimate accuracy, the MMP technique is around 200 times faster. It is advised to increase the speed and accuracy by using CSI information from multiple packets using a multipacket CSI aggregation technique. A novel device-free localization system called *DeFi* was developed by Ref. [29], which employs an AoA and equivalent ToF measurement-based background reduction algorithm to eliminate static paths and distinguish the target reflection path from motion paths. DeFi uses a particle filter technique to estimate the target location while using the AoA of the target reflection path as observation data. DeFi's median errors in controlled conditions are 0.57 m, while its median errors in corridor environment are 0.34 m.

2.2 Fingerprinting-based indoor localization approach

Fingerprint localization is an approach for correlating particular locations within a room with acquired signal features. It has several advantages over AoA and ToA in terms of ease of implementation, accuracy, and algorithmic flexibility. In the fingerprint localization system, Wi-Fi is often employed as a signal source. Two phases are involved in this type of positioning: offline and online. The system builds a fingerprint database when it is offline by collecting data from various locations. The system compares the collected signal characteristics with those in the database during the online phase. The schematic diagram of the process is shown in Fig. 3.

2.2.1 Fingerprinting preprocessing

To extract more precise fingerprint features, it is necessary to have more precise location information. CSI gives you information on both amplitude and phase unlike traditional RSSI, which only offers one dimension of data. Chen et al. [30] created fingerprints using RSSI, transmission power, and CSI. However, in crowded indoor environments, 90% of the generated detection spots will be accurate to within 2 m. The author of Ref. [31] proposed a novel method called *FapFi* for passive human indoor localization. This method incorporates CSI amplitude and phase information for an average positional accuracy of 0.5–1.2 m in labs and conference rooms. After the

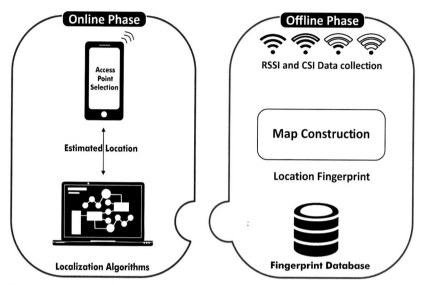

Fig. 3 Schematic block of the fingerprint positioning principle.

offline phase's removal of duplicate values and outliers, the calibrated phase data are extracted using a linear transformation. Accurate location estimate and fingerprint database enhancement are made possible by the high dimensionality of the CSI sample matrix that was collected. A problem arises, though, due to the fingerprint database's high dimensionality and temporal complexity. Shi et al. [32] suggested principal component analysis or linear decision analysis as a solution to this problem in order to choose the most relevant feature vectors and reduce the number of dimensions.

2.2.2 Fingerprinting matching

The CSI fingerprint location's online matching phase can use either the probabilistic or the deterministic algorithm. Wu et al. [33] presented a method for localization using augmented naive Bayesian classifiers that have increased accuracy and could achieve a position accuracy of 2 m. Similarly, Shi et al. [34] suggested a fingerprint-based passive indoor position monitoring system that is device-free and produces fingerprint samples using CSI amplitude. They used Bayesian filtering in the online stage to recursively estimate the coordinates of moving objects in order to monitor subjects' movement while they are in crowded areas, where the system's positioning accuracy is less than 1 m.

2.2.3 Fingerprinting using CSI

Recently, the benefits of CSI fingerprint positioning technology, including its high accuracy, low cost, algorithmic flexibility, and ease of installation, have led to the emergence of several CSI fingerprint positioning systems. Xiao et al. [35] developed the CSI-based *FIFS* indoor fingerprint localization system, where frequency and spatial diversity cause the initial CSI value to change. Afterward, it is converted into a fingerprint using a probability model, which is then saved in a database. It is less accurate than the Horus system when installing an unaltered *FIFS* WLAN network, with an average error of less than 1 m. The advantages of frequency diversity are not completely achieved even if *FIFS* just employs the amplitude features of CSI and averages the amplitudes measured on multiple antennas to build fingerprints.

The use of the CSI fingerprint-based positioning approach has grown significantly due to its high level of accuracy. The training time and complexity of the processing both increased in parallel with the database size. The extraction of CSI data attributes and training data using deep learning approaches can improve the speed and accuracy of fingerprint acquisition and matching. In the offline phase when time and frequency synchronization problems are resolved, the CSI of the 10 MHz bands is analyzed and combined to create 1 GHz fingerprints. Afterward, when a database stores by combining these fingerprints with the test point fingerprint to predict the location, the Tx resonance strength is calculated during the online phase. In an NLoS office environment, this system can achieve an accuracy of 5 cm on a small area of 20×70 cm^2 using one antenna. Zhang introduced a single access point positioning system that used a linear transform technique to get rid of phase noise and a phase decomposition method to collect multipath phase data. The system's average accuracy was 0.6 m in the lab, 0.45 m in the conference room, and 1.08 m in the corridor. Similarly, Khan et al. [36] proposed a DFL approach using CSI fingerprint samples and universal software-defined radio peripheral devices to identify and locate a single human subject participating in activities. By reducing dimensions and utilizing PCA for the optimal feature selection, they explored to locate six distinct activities in a 3.8×5.2 m^2 office environment setup block diagram, which is shown in Fig. 4.

Comparisons of seven ILSs are provided, as shown in Table 1. Laboratory, corridor, conference room, and living room are among the common indoor environments.

Fig. 4 Diagram illustrates horizontal and vertical zones.

Table 1 System comparison for indoor localization.

System	Method	CSI information	Experimental setup and accuracy
FIFS [35]	Fingerprinting	Phase and amplitude	Lab and corridor, up to 1 m
FILA [23]	Range	Amplitude	Lab, library, and corridor, up to 3 m
PhaseFi [37]	Angle and range	Phase	Lab, room, up to 2 m
ConFi [38]	Fingerprinting	Amplitude	Lab, room, up to 1.37 m
LiFS [25]	Range	Amplitude	LoS with 0.5 m, NLoS with 1.1 m
S-Phaser [26]	Range	Phase and amplitude	Lab 1.5 m
DeFi [30]	Angle and range	Phase	Lab, room, 0.6 m
ResLoc [39]	Angle	Phase and amplitude	Lab, corridor, 0.89–1.24 m

3. Technology-based indoor localization techniques

Data transmission between the transmitter and the receiver sections of ILS requires methods of communication. These enabling technologies can be classified according to a number of factors, including the signal's type and range. There are a few ways to classify the signals.

1. Radio frequency: Specifically, frequencies between 2 and 5 GHz, such as Wi-Fi.
2. Light and optics: They function similar to visible and infrared light.
3. Sound: There are also acoustic and ultrasonic sounds.
4. Additional types, such as magnetic.

The range criteria can be used to divide these technologies into short and long ranges [40]. The short-range technologies can be used in small indoor spaces due to their range, which ranges from a few centimeters to a few hundred meters [41]. Long-range technologies can communicate with thousands of devices at a distance of only a few hundred meters. Low-power wide-area networks (LPWANs), such as Sigfox and LoRaWAN, which are the two leaders outside of the cellular industry, are not feasible localization methods due to the multipath and fading effects of the signal over long distances. However, they could provide a accurate result in enclosed large arenas when combined with indoor technology. The main use cases are large indoor or covered venues such as mega malls, college campuses, and smart parking. In the section, we will examine and compare a few of the most well-known technologies.

3.1 Wi-Fi

WiFi is a popular communication method that utilizes the ISM band at frequencies of 2.4 and 5 GHz, with its 802.11ax variation operating in the 1–6 GHz band. For IoT applications, it has been updated to Wi-Fi 6 or 802.11ax and conforms to the IEEE 802.11 standard. In areas with a high device density, this new version offers four times quicker speeds and less energy usage. The Wi-Fi alliance has included a new RTT function to the 802.11.mc standard that allows precise localization within a range of 1–2 m. This function calculates the duration of time a signal needs to travel to the access point [42] by triangulating the location using data from several access points. The accuracy of signals in noise reduction methods is significantly impacted by interference in the ISM band of communications. Although in [43] that accuracy below 1 m is achievable, commercial

Fig. 5 Wifi-based indoor localization.

implementations often achieve an accuracy of around 2 m, dependent on the surroundings and object movement [44,45]. Numerous machine learning algorithms, such as k-nearest neighbor, artificial neural network (ANN), support vector machine (SVM), k-means, and random forest (RF), have been applied for pattern matching during the online phase in order to reduce the adverse impacts of RSSI fluctuations. This method of localization cannot be as accurate depending on the Wi-Fi connection's strength. Fig. 5 depicts the schematic of the Wi-Fi-based localization method.

3.2 Radio frequency identification (RFID)

RFID is a pioneering technology used in supermarkets and warehouses to track the arrival and departure of goods. It transmits data using electromagnetic impulses. Reader, tag, and server make up the three parts of an RFID system. There are three subcategories in each of the three categories: active, passive, and semipassive. Passive RFID can work up to 10 m. Active RFID tags operate as beacons, emitting data up to 100 m away while operating in the UHF and microwave frequency bands. These are suitable choices for inexpensively tracking an object at a short distance [46]. RFID localization accuracy and range are influenced by a number of factors, including tag and reader density, antenna type, and frequency [47]. RFID is not the best choice for tracking and navigation systems while being widely used as a cheap and simple positioning technology. Active RFIDs have been utilized for tracking in certain situations, such as the LANDMARC system presented by Ref. [48], which has a large range and is energy-efficient but has problems with tracking latency. Another example is the real-time RFID indoor positioning system described by Ref. [49], which uses Kalman filters to minimize drift and Heron bilateration to estimate the location. In addition, Siachalou et al. [50] presented an RFID-based phased fingerprint-based locating system for monitoring in sizable warehouses and retail stores. Fig. 6 provides a graphic depiction of RFID-based localization.

Fig. 6 RFID-based indoor localization.

3.3 Ultrawide band (UWB)

UWB technology has become a competitive alternative for communication in an ILS due to its low power consumption, high data transmission rates (up to 1 Gb), and high accuracy. Although operating in the 3.1–10.6 GHz frequency spectrum, UWB encounters very little interference from other bands. There are four specific data transmission speeds inside UWB: 110 Kbps, 850 Kbps, 6.8 Mbps, and 27 Mbps. RTT, TDoA, and ToA are used by UWB to accurately determine the distance between tags and anchors as compared to RSSI techniques. The advantages of UWB over traditional narrow-band systems are high security, minimum complexity, low power consumption, resistance to multipath effects, and high accuracy. In commercial products, accuracy can be achieved to a range of roughly 20 cm [51]. In order to provide indoor and outdoor localization solutions, the *EIGER* project combines GNSS and UWB technologies globally. For precise ILS, such as tracking the movements of robots and spotting troops on the battlefield, UWB technology is particularly beneficial. UWB modules are recommended for industrial IoT use cases like warehouses since they are economical for usage in large-scale applications. In Ref. [52], a UWB system that does not need previous information and can position appropriately in challenging situations is proposed. The authors were successful in lowering the absolute range errors' root mean square following NLOS mitigation from the initial 1.3 to 0.651 m in their experiment, which was carried out in a real office setting. The localization based on UWB is shown in Fig. 7.

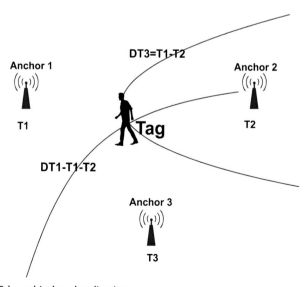

Fig. 7 UWB-based indoor localization.

3.4 Vision

Werner et al. [53] suggest that the prevalence of surveillance cameras in indoor environments makes the usage of indoor monitoring systems in public areas like shopping malls a cost-effective security solution. In this scenario, the camera on a smartphone can serve as a sensor for computer vision, allowing camera-based localization within the building. This method combines image identification with distance approximation to estimate the object's coordinates. In Ref. [54], Chen et al. suggested the use of an optical camera and an alignment sensor node, wherein nearby nodes are identified using fingerprints from the Wi-Fi signal. An enhanced localization technique [55] includes capture coordinates and image information needed to structure fingerprints. This approach works effectively in both indoor and outdoor environments, as well as in shadowed areas. There are numerous novel algorithms that solve the distortion and accuracy problems with vision-based systems. This method requires a lot of computing power and is inefficient in crowded environments [56]. The vision-based localization for multiple humans using a low-cost vision sensor is shown in Fig. 8.

Fig. 8 Vision-based indoor localization.

3.5 Bluetooth

Bluetooth is a wireless technology that uses a low power consumption and operates on the 2.4 GHz ISM band. The version 4.2 range is up to 70–100 m, while version 5.0's range is up to 400 m at a speed of 24 Mbps. Batteries are the power supply for Bluetooth beacons, which are compact, affordable devices. If a user is within the range of many beacons, they are able to roughly calculate the distance to the beacon, which may be used to estimate their indoor location. Bluetooth beacons are a desirable solution for location-based applications due to these features [57,58]. Bluetooth v5.1 provides a precise and reliable indoor localization capability without the requirement for expensive hardware or an LoS. This feature is used to calculate the angle of the Bluetooth signal using two widely used methods: AoA and AoD. The *orange dot* in Fig. 9 represents the estimated position of a mobile tag.

Precision of 1–2 m may now be achieved by industrial devices, which is suitable for tracking and navigation systems for large objects and people [59]. At a short range, Bluetooth is more accurate than Wi-Fi, while Wi-Fi is more accurate and dependable in wider spaces [60]. Bluetooth and Wi-Fi both operate on the 2.4 GHz frequency band, but despite this overlap, the two technologies do not interfere with each other's signals. A combination of Bluetooth and Wi-Fi performance can enhance

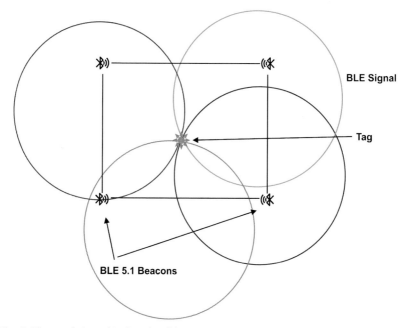

Fig. 9 Bluetooth-based indoor localization.

positioning systems, as stated in Ref. [61]. Bluetooth low energy (BLE) technology, known for its low energy consumption and high accuracy, is a competitive technology. The author of Ref. [52] used a simple trilateration
method based on RSSI to assess the effectiveness of Wi-Fi, BLE, and
ZigBee. They observed that the methods had a respective accuracy of
48.5 cm, 84.4 cm, and 91.1 cm respectively.

3.6 LorawAN

LoRa is renowned for its resilience to Doppler, multipath, and interference
effects. Two strategies are offered by LoRaWan: TDoA for fine accuracy
and RSSI for a course [62]. According to the author's of [63], LoRa is more
trustworthy and can tolerate harder environments than Bluetooth and
Wi-Fi. As a consequence, using noise filters like Kalman does not considerably improve the accuracy as compared to Bluetooth and Wi-Fi. LoRa
can reach mean errors of 1.19 and 1.72 m in LoS and non-LoS environments, respectively. LoRa is intended for large-scale indoor and outdoor
localization scenarios [64], and it reduces non-Gaussian and Gaussian noise
in RSSI-based methods. The Lora-based localization is shown in Fig. 10.

Remarks: In Table 2, overview of the current underlying technologies
demonstrates that submeter accuracy is achievable. The three primary metrics
for contactless activity application considerations are cost, energy, and
accuracy. The most promising technologies in this comparison are UWB
and BLE. The most popular UWB is now employed in industrial applications

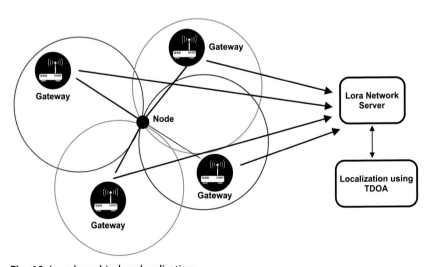

Fig. 10 Lora-based indoor localization.

Table 2 Range-wise technologies comparisons for indoor localization.

Technology	Technique	Cost	Pros	Cons	Accuracy (m)	Coverage (m)	Spec
WiFi	CSI, RSSI, Proximately	High	Doesn't need LoS, has good accuracy and is accessible on most smartphones	Mobility, multipath effect, LoS problem	1–6	Up to 100	WiFi-5, WiFi-6
RFID	RSSI, TOA, TDOA, AoA	Low	Detects multiple tags quickly and simultaneously without needing LoS between Tx and Rx	Passive tags' capabilities are restricted by their small coverage, signal fluctuation, and multipath effect	Passive 1 m and active 5 m	Up to 100	Active, passive and semipassive
UWB	TDOA	High	Penetration through walls and other obstructions, great accuracy, and low energy consumption	Performance degrades in NLoS due to limitations in expensive coverage	0.2–2.0	Up to 200	Available in iPhone 11 onward
Vision	Scene analysis	Medium	Fusion of image data and information from other sensors may increase the performance	Additional depth information is necessary for the transformation from image space to object space	0.1	LoS	Shopping mall and public places
Bluetooth	RSSI, TDOA	Low–medium	Good accuracy does not need LoS, and available in most devices	Mobility, RF interference, and insufficientcoverage	3–12	Up to 400	BLE 5.1 for indoor
LoraWAN		Low	Open standard. These are ideal for long-distance crops without access to the mains or batteries	The biggest drawback is the need for local networks of this type and providing coverage	2–6	5 km in urban, 20 km in rural	ECID and 5G

like smart industries since it is accurate but not cost-effective. However, by lowering the cost and adding chips to devices like the iPhone, it hopes to increase user availability.

4. Machine learning for indoor localization

In recent years, indoor localization has made significant advances due to the use of AI/ML techniques [65,66]. The core advantage of AI/ML is their capacity to use observable data to effectively make decisions without the need for precise mathematical formulations. As ML techniques are essential in the field of indoor localization, this section examines the current algorithms being utilized in the literature to achieve this objective.

The study examines different aspects of the results, including the algorithms used, signal types, number of access points and reference points, metrics, experimental methods, and the radio maps commonly used for indoor automated localization algorithms. Fig. 11 shows the distribution of algorithms, and it indicates that ANN-based algorithms are the most frequently used, with approximately 70% of the analyzed papers using ANN or one of its variations. These NN-based methods are effective for Wi-Fi

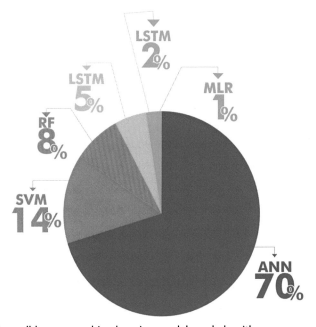

Fig. 11 The well-known machine learning models and algorithms.

signals that are subject to fluctuations due to their nonlinear functions. In recent years, there has been a growing trend toward ANN-based solutions, with ANN being used in 20 out of 24 publications by 2021. The most promising results were achieved using a deep neural network to process Wi-Fi mmWave signals, with an average error of 0.11 m [67]. However, these results are not conclusive due to various factors that may influence them, such as the size of the test area and the number of access point used.

We explore more into the NN, SVM, and RF specifications in the sections that follow.

4.1 Neural networks (NN)

The NNs consist of three layers of interconnected nodes: input, one or more hidden, and output. During the training phase, the model's results are evaluated against the desired result, and the resulting error is calculated. As an error propagates through the hidden layers, the nodes' weights are adjusted to boost performance. This iterative process is continued in order to enhance the model's performance since there is not a single distinct configuration that has been explicitly discussed in the literature. Researchers continue to experiment until they find a satisfactory balance between accuracy and computation time. Roy et al. [68] achieves 2.4 m accuracy with six layers of 512 nodes each. Zhang et al. [69] employ four levels without specifying the number of nodes, whereas Liu et al. [70] use two layers of 50 nodes each. Thus, there is no standard configuration. However, the dilemma of choosing the right configuration varies depending on a multitude of factors, including the scenario's nature, shape, presence of obstacles, the number of used access points, and other factors.

4.2 Support vector machine (SVM)

The SVM algorithm family divides results into two sections on a plane and classifies them accordingly. Christy et al. [71] achieved a 2.7 m accuracy using the M-LS-SVM algorithm, which differs from the original SVM by utilizing linear functions instead of quadratic functions. Similarly, Schmidt [72] acheived a 0.7 m accuracy using the SVM classifier directly in a comparable environment. In contrast, Yin et al. [73] utilized an SVM algorithm employing CSI instead of RSSI, achieving a 1.909 m accuracy in a simpler scenario without rooms or obstructions.

4.3 Random forest (RF)

The RF algorithms prioritize the creation of several decision trees in order to construct a learning model. The result is eventually determined by the classification that is most common among the decision trees, which contributes to the overall model. To the best of our literature study, in 2021, no papers were found that utilized it due to a decline in its indoor positioning use. Only three papers utilized it in 2020, and none of them conducted experiments using scenarios with obstacles. The best RF accuracy measured in 2020 in a $112m^2$ area is 1.68, according to Ref. [74]. Liu et al. [75] reported an accuracy of 1.20 m within a $75m^2$ area in contrast to Ref.'s [76] writers, who demonstrated an accuracy of 0.4033 m within an 80 m^2 area in 2018. These results suggest that RF is a viable approach for small-scale applications.

Remarks: Choosing the appropriate algorithm is dependent on various factors including the computing power that is accessible, the volume of data that needs to be analyzed, and the type of the infrastructure (walls, tables, rooms, etc.) required to deploy the fingerprint-based system. In comparison to SVM, RF methods use less computational resources. SVM-based approaches are not appropriate for large datasets since the training complexity of SVM is inversely related to the size of the dataset. SVM algorithms are effective in lowering the non-LOS of signals at the level of infrastructure complexity, making them suitable for small and complex environments. The number of nodes and neurons in a network can be changed to optimize speed, whereas NNs are more programmable and can be modified for increased performance based on lower accuracy. Being specially tuned decision trees that can easily handle massive volumes of data, RFs, on the other hand, are well suited for clustering models in larger scenarios. ANNs are the method of choice for dealing with background noise and interference from various paths and several access points. Therefore, SVM is the best alternative for constrained environments with limited processing resources, while RFs and NNs are more difficult to deploy in complex environments with large datasets but more versatile owing to their vast configuration capabilities.

5. Discussion and challenges

In this section, we address the key implementation challenges and possible solutions for indoor localization.

5.1 Evaluation metrics

Metrics for evaluation highlight the factors that could influence a system's effectiveness and performance. Several of these precautions are very important due to the limitations of IoT devices. Here are definitions of the most prevalent metrics used in this chapter.

1. **Accuracy:** Accuracy is how closely a measurement or estimate reflects the actual value. Numerous factors, including signal transmission, LoS, noise, and others, affect it. It is a core part of ILS, and the majority of research focuses on increasing accuracy by utilizing noise filters to eliminate signal noise and system errors.

2. **Precision:** Precision is the constancy of a number through time and between measurements, as contrasted to accuracy, which is how closely a measurement reflects the real value. It implies that the various tests' results are scattered fairly equally. A test is considered as valid when it is both accurate and precise.

3. **Availability:** Availability defines service and device availability. Users do not need proprietary hardware to utilize it. Wi-Fi and Bluetooth exist in almost all smartphones, unlike UWB. Visible light communication is a popular alternative. Availability indicates that a service is available when needed. System downtime is deducted from agreed-service time.

4. **Cost:** The cost of localization systems influences their commercial viability. The issue is complex and goes beyond hardware. While implementing a product, time and human resources are essential. Existing expenditures like energy consumption and maintenance are out of reach for many small businesses. The user, maintenance, or current expenses of an ideal localization system should not be excessive.

5. **Coverage:** A cost-efficient industrial localization system covers large areas with the least nodes. Increasing ranges may degrade performance. Tx-power determines the system's use case coverage.

6. **Energy Efficiency:** It is essential for portable nodes that are worried about battery life and drain. Precision, latency, and energy all have trade-offs. The advertising interval and Tx-power could have an influence on energy consumption. The time between sending packets is the advance for mobile things that move quickly. Although less effective than advertising, it uses more energy. It is an essential step for IoT devices with constrained energy.

7. **Latency:** The time gap between a request and a location is called a delay or latency. In order to be effective, real-time localization systems must be

capable of locating an object in less than 1 ms and lose only a few operations. It is difficult to work hard in real time due to the noise in the
vicinity. Time-sensitive services like smart hospitals and IoT healthcare
emergency systems are impacted by latency.

8. **Robustness:** The capacity of a system to tolerate disturbances and signal
losses that affect its operations. Environment-related mistakes must be
handled via a robust positioning system. This precaution guarantees that
all systems operate consistently, particularly in commercial or emergency
situations.

5.2 Possible solutions

In this section, we explore at the implementation challenges as well as
potential solutions and difficulties.

1. **Cost:** A preliminary estimate was needed since the cost is the key factor
influencing all projects. Anchors, servers, specialized software, and other
hardware and infrastructure are included in the direct costs. Two other
expenses that are not included are installation and maintenance costs.
One such case is the application of fingerprints, which requires too much
time and resources. Due to the expenses incurred associated with indoor
localization, such as energy usage, battery use, and monthly expenses,
users also hesitate to accept it. Hardware prices may remain low if
consumers choose low-cost mobile options like Bluetooth and Wi-Fi.

2. **Energy efficacy:** Indoor localization often has the problem of energy utilization that affects indoor localization. Energy utilization, accuracy, and
performance are challenging to balance. Accurate systems need strong,
consistent signals, which drain batteries. Battery life affects mobile-node-
based localization (MNL) user satisfaction. The configurations of MNLs,
the use of task-management systems, and the prevalence of electronic
communication systems are alternative solutions. First, passive and active
Bluetooth and UWB (in TDoA) nodes are employed. Using an algorithm
or outsourcing may help with location. User devices outsource positioning tasks to servers to save energy; however, this delays services.
Lightweight algorithms save energy but reduce accuracy and security.
Other options include adjusting the advertisement time and Tx power.

3. **Noise and obstacles:** Obstacles and noise make indoor localization
difficult. Multipath is caused by distraction, fading, and reflection.
Multipath signals are NLoS, making distance, strength, and angle difficult to determine. Noise affects calculations and lab prototype test results.

Two solutions are offered that is manipulating signals by reducing ambient noise and obstacles. In real-world circumstances, we need algorithms to eliminate external environmental barriers and noise. The median, Kalman, and particle filters are suggested in the literature. The simplest filters for removing noise are median and average filters. These filters maintain signal patterns even under changing environments. When there is Gaussian noise in a linear system, Kalman filters (KF) are often used. Nonlinear functions could be resolved via non-Gaussian propagation. The Taylor expansion is used to approximate extended KF (EKF) linearly. High nonlinear systems delay or cause EKF to diverge (over first- and second-order estimates). As a result, UKF is increasing the number of particles around with a point, which will increase the accuracy and reliability of the particle filter (PF). In order to reduce noise, PF employs Monte Carlo iterations. It is more accurate, precise, and noise-resistant while being faster. A number of accurate methods and filters have been given to address this problem, which requires extensive processing. As a result, energy-constrained hardware, such as mobile phones, necessitates the development of light-weight algorithms with low computational and energy consumption.

4. **Security and privacy:** Commercial providers face security and privacy issues while placing services. Researchers prioritize accuracy above security. Location data endanger consumers' privacy and security. (i) Database corruption (ii) RF interference (iii) malicious nodes (iv) IoT privacy standards, and (v) network security are important indoor localization threats. ISM-based communications are more susceptible to RF interference than licensed channels. Developers cannot use complex algorithms to handle security and privacy issues due to energy and processing constraints. Some lightweight techniques find phoney malicious nodes and offer system privacy.

5. **Reliability:** It is a problem with commercial goods, especially emergency systems. When services fall down due to a natural disaster and there is no electricity or Internet, it is vital. When the system experiences unforeseen circumstances, a backup solution must be taken into account.

6. **Lack of interoperability and Standards:** Many companies provide products based on their own proprietary ecosystem. Without a thorough standard, regulating and regulating this business is difficult. Some businesses combined Bluetooth, UWB, and Wi-Fi to increase interoperability. There is no open location tracking standard that can meet all IIoT application and user requirements. IoT-based systems

and location-based services use a wide range of communication protocols and heterogeneous hardware. The interaction of several components is a significant issue in this field. This issue may be solved by developing interoperable middle ware using standard tools and protocols. A great alternative could be to create a single hub that uses all communication technologies. Due to the uneven distribution of products and communication technologies, using edge technologies like fog may be a wise decision. Specialized fog nodes may detect the traffic of objects based on their nature and communication technology.

To overcome challenges and increase the effectiveness of the ILS, there are several techniques and strategies:

1. **Software:** One of these tools that may help in two different ways is an algorithm. The noise and error in systems must first be reduced. Using machine learning, we can further analyze and predict how things will behave. Recently, deep learning algorithms (CNN, RNN) have outperformed well-established neural networks, SVM, and KNN in terms of effectiveness.

2. **Hardware acceleration:** It is better than software when it involves task completion. In addition to using less resources, it reduces noise and improves accuracy.

3. **Design and management:** Unwanted events may be avoided and resource waste can be reduced with appropriate design and management. Academic papers address the accuracy problem. In indoor applications, noise, energy consumption, security, and privacy are significant obstacles. Costs and dependability make sense for commercial products.

5.3 Companies and brands

Indoor locations including smart buildings, hospitals, and factories are in greater demand. According to statistics from the ResearchAndMarket website, the global indoor market is anticipated to reach a value of 23.6 billion by 2025, up from an estimated 6.92 billion in 2020. We note that several major businesses including Microsoft, Google, and Apple have shown various solutions in this field [77], which is indicative of the positioning services market's promise. The GPS's limitations are being addressed via indoor solutions, but their accuracy increased from meters to centimeters. Indoor has several applications in the business, government, retail, health, military wearable device, and inventory monitoring fields. Numerous businesses have joined this industry and adapted their goods with location-based services due to the limitless uses and strong demand for localization services. For instance,

Table 3 Manufacturers solution and indoor vendors.

References	Technology	Vendor	Manufacturers	Usage
[78]	Bluetooth, UWB, WiFi, LTE-M	Estimote	Both	Hall and museum
[79]	Bluetooth, UWB, WiFi, RFID	Infsoft	Both	All fields
[79]	Bluetooth, UWB, WiFi, RFID	Infsoft	Both	All fields
[80]	Bluetooth, UWB, WiFi, RFID	Inpixon	Both	Shopping mall, hospital
[81]	Bluetooth, Lora, WiFi	MOKOSmart	Both	Smart home, fitness room
[82]	RFID reader, tag, antenna	Zebra	Manufacturer	People tracking and tracking

Bluetooth Sig introduced BLE 5.1, which had a specific location feature. Table 3 lists a few of the current vendors along with the technologies and application cases they support. We examine some of these vendors in the section below.

1. **Estimote:** It is a leading brand for localization use cases, especially in proximity-based solutions, due to a programmable indoor prototyping platform. Estimote beacons support BLE, WiFi, and UWB technologies. It is one of the first companies to provide approximation services in retail and workplaces, according to Ref. [78].

2. **Infsoft:** Infsoft provides indoor digitalization, navigation, tracking, and analytic solutions using a range of technologies, including BLE, UWB, WiFi, RFID, and cameras. The hardware includes transmitter tags (UWB, BLW, and LoRa) and receiver nodes (Infsoft locator). It works with BLE versions 4.0 and 5.1. The locating node has the ability to handle BLE, UWB, and WiFi sensors all at once. Infsoft offers solutions for practically all fields, according to its extensive range of real-world use cases [79].

3. **Inpixon:** This company is involved in data collection, analysis, and visualization to make space smart and safe. It benefits from sensor fusion, position analytics, and the IoT. It makes use of a range of radio frequency (RF) technologies, including cellular, Wi-Fi, BLE, and UWB. It works with businesses and public institutions, including governmental offices, corporate campuses, medical facilities, and shopping centers [80].

4. **MOKOSmart:** It is a company that specializes in creating and manufacturing IoT products for smart devices. In IoT devices and gateways, it uses BLE, WiFi, and LoRaWAN. It comprises a wide range of applications, such as smart agriculture, indoor and outdoor solutions, asset monitoring, smart retail, and asset management, and joined the LoRa Alliance [81].

Indoor systems have many active vendors, and this industry is expanding as a result of strong demand. Since we are unable to cover the whole market in one article, we have chosen a few of the more well-known ones. Unfortunately, the scale and type of use case directly affect how much the products are estimated to be worth.

5.4 Conclusion and future recommendations

The main focus of this chapter was to investigate the various localization methods used to improve the accuracy of identifying elderly people's activities at home or in healthcare environment. Accurate location information is necessary for support services to be effective and to enhance the lives of the elderly. Nevertheless, there have been a few studies that combine indoor location data with system for identifying daily activities. Therefore, we highlighted the necessity of indoor localization services, as GPS alone cannot cover all indoor spaces. By integrating GPS with indoor localization, a comprehensive location-based system that works in both inside and outdoor environments can be developed. This chapter provides a comprehensive overview and comparison of the recent approaches, methods, and technologies for indoor localization. The outcomes of different approaches can help prevent incidents, predict human behavior for future planning, and ensure a satisfactory user experience.

Section 5 addresses the key challenges that occur while deploying location-based services in IoT applications. There are still a lot of risks and opportunities in this field, which calls for interested parties to do additional in-depth and specialized studies. Here, it is necessary to address the lack of a consistent and comprehensive platform that combines various technologies for performance evaluation and data analysis. To achieve this objective, we require a hybrid system with high accuracy, low cost, and adequate coverage of long- and short-range communication modules. Several indoor options such as Bluetooth 5.1, UWB, VLC, LoRa, RFID and cellular are available, which will help the researchers to choose the best technology and approach. However, it is important to note that the results presented in this chapter are not enough for decision-making and can only be valid when conducted in the same testbed under standard density, size, and ambient noise conditions. Lastly, this chapter highlights the challenges associated with indoor positioning and recommends some potential directions for this rapidly evolving domain. Overall, this chapter provides a useful manual for comprehending, applying, and developing approaches employed in different scenarios.

References

[1] S.-J. Yoo, S.-H. Choi, Indoor AR navigation and emergency evacuation system based on machine learning and IoT technologies, IEEE Internet Things J. 9 (2022) 20853–20868.

[2] M.Z. Khan, A. Taha, W. Taylor, A. Alomainy, M.A. Imran, Q.H. Abbasi, Indoor localization using software defined radio: a non-invasive approach, in: 2022 3rd URSI Atlantic and Asia Pacific Radio Science Meeting (AT-AP-RASC), IEEE, 2022, pp. 1–4.

[3] P. Kanakaraja, et al., IoT enabled BLE and LoRa based indoor localization without GPS, Turk. J. Comput. Math. Educ. (TURCOMAT) 12 (4) (2021) 1637–1651.

[4] S. Capolongo, M. Gola, A. Brambilla, A. Morganti, E.I. Mosca, P. Barach, COVID-19 and healthcare facilities: a decalogue of design strategies for resilient hospitals, Acta Biomed. 91 (9-S) (2020) 50.

[5] A.N. Sert, Theory and practice in social sciences, in: Senior Tourism in the Aging World, St. Kliment Ohridski University Press, pp. 488–498.

[6] J. Wyffels, J.-P. Goemaere, P. Verhoeve, P. Crombez, B. Nauwelaers, L. De Strycker, A novel indoor localization system for healthcare environments, in: 2012 25th IEEE International Symposium on Computer-Based Medical Systems (CBMS), IEEE, 2012, pp. 1–6.

[7] T. Van Haute, E. De Poorter, P. Crombez, F. Lemic, V. Handziski, N. Wirström, A. Wolisz, T. Voigt, I. Moerman, Performance analysis of multiple indoor positioning systems in a healthcare environment, Int. J. Health Geogr. 15 (1) (2016) 1–15.

[8] C.S. Leung, J. Sum, H.C. So, A.G. Constantinides, F.K.W. Chan, Lagrange programming neural networks for time-of-arrival-based source localization, Neural Comput. Appl. 24 (1) (2014) 109–116.

[9] F. Erden, S. Velipasalar, A.Z. Alkar, A.E. Cetin, Sensors in assisted living: a survey of signal and image processing methods, IEEE Signal Process. Mag. 33 (2) (2016) 36–44.

[10] J. Wahlström, I. Skog, Fifteen years of progress at zero velocity: a review, IEEE Sensors J. 21 (2) (2020) 1139–1151.

[11] R. Ramirez, C.-Y. Huang, C.-A. Liao, P.-T. Lin, H.-W. Lin, S.-H. Liang, A practice of BLE RSSI measurement for indoor positioning, Sensors 21 (15) (2021) 5181.

[12] M. Afzalan, F. Jazizadeh, Indoor positioning based on visible light communication: a performance-based survey of real-world prototypes, ACM Comput. Surv. (CSUR) 52 (2) (2019) 1–36.

[13] J.A. Paredes, F.J. Álvarez, T. Aguilera, J.M. Villadangos, 3D indoor positioning of UAVs with spread spectrum ultrasound and time-of-flight cameras, Sensors 18 (1) (2017) 89.

[14] K. Pahlavan, P. Krishnamurthy, Evolution and impact of Wi-Fi technology and applications: a historical perspective, Int. J. Wireless Inf. Networks 28 (1) (2021) 3–19.

[15] Á. De-La-Llana-Calvo, J.-L. Lázaro-Galilea, A. Gardel-Vicente, D. Rodríguez-Navarro, B. Rubiano-Muriel, I. Bravo-Muñoz, Analysis of multiple-access discrimination techniques for the development of a PSD-based VLP system, Sensors 20 (6) (2020) 1717.

[16] A. Cidronali, S. Maddio, G. Giorgetti, G. Manes, Analysis and performance of a smart antenna for 2.45-GHz single-anchor indoor positioning, IEEE Trans. Microwave Theory Tech. 58 (1) (2009) 21–31.

[17] Y. Wang, S. Ma, C.L.P. Chen, TOA-based passive localization in quasi-synchronous networks, IEEE Commun. Lett. 18 (4) (2014) 592–595.

[18] N. Lasla, A. Bachir, M. Younis, Area-based vs. multilateration localization: a comparative study of estimated position error, in: 2017 13th International Wireless Communications and Mobile Computing Conference (IWCMC), IEEE, 2017, pp. 1138–1143.

[19] R. Schmidt, Multiple emitter location and signal parameter estimation, IEEE Trans. Antennas Propag. 34 (3) (1986) 276–280.

[20] J. Xiong, K. Jamieson, ArrayTrack: a Fine-Grained indoor location system, in: 10th USENIX Symposium on Networked Systems Design and Implementation (NSDI 13), 2013, pp. 71–84.

[21] J. Gjengset, J. Xiong, G. McPhillips, K. Jamieson, Phaser: enabling phased array signal processing on commodity WiFi access points, in: Proceedings of the 20th Annual International Conference on Mobile Computing and Networking, 2014, pp. 153–164.

[22] L. Zhang, H. Wang, 3D-WiFi: 3D localization with commodity WiFi, IEEE Sensors J. 19 (13) (2019) 5141–5152.

[23] K. Wu, J. Xiao, Y. Yi, M. Gao, L.M. Ni, FILA: fine-grained indoor localization, in: 2012 Proceedings IEEE INFOCOM, IEEE, 2012, pp. 2210–2218.

[24] M. Kotaru, S. Katti, Position tracking for virtual reality using commodity WiFi, in: Proceedings of the IEEE Conference on Computer Vision and Pattern Recognition, 2017, pp. 68–78.

[25] J. Wang, J. Xiong, H. Jiang, K. Jamieson, X. Chen, D. Fang, C. Wang, Low human-effort, device-free localization with fine-grained subcarrier information, IEEE Trans. Mob. Comput. 17 (11) (2018) 2550–2563.

[26] S. Han, Y. Li, W. Meng, C. Li, T. Liu, Y. Zhang, Indoor localization with a single Wi-Fi access point based on OFDM-MIMO, IEEE Syst. J. 13 (1) (2018) 964–972.

[27] S. Han, Y. Li, W. Meng, C. He, A new high precise indoor localization approach using single access point, in: GLOBECOM 2017-2017 IEEE Global Communications Conference, IEEE, 2017, pp. 1–5.

[28] A.U. Ahmed, R. Arablouei, F. De Hoog, B. Kusy, R. Jurdak, N. Bergmann, Estimating angle-of-arrival and time-of-flight for multipath components using WiFi channel state information, Sensors 18 (6) (2018) 1753.

[29] L. Zhang, Q. Gao, X. Ma, J. Wang, T. Yang, H. Wang, DeFi: robust training-free device-free wireless localization with WiFi, IEEE Trans. Veh. Technol. 67 (9) (2018) 8822–8831.

[30] D. Chen, L. Du, Z. Jiang, W. Xi, J. Han, K. Zhao, J. Zhao, Z. Wang, R. Li, A fine-grained indoor localization using multidimensional Wi-Fi fingerprinting, in: 2014 20th IEEE International Conference on Parallel and Distributed Systems (ICPADS), IEEE, 2014, pp. 494–501.

[31] X. Dang, X. Si, Z. Hao, Y. Huang, A novel passive indoor localization method by fusion CSI amplitude and phase information, Sensors 19 (4) (2019) 875.

[32] S. Shi, S. Sigg, L. Chen, Y. Ji, Accurate location tracking from CSI-based passive device-free probabilistic fingerprinting, IEEE Trans. Veh. Technol. 67 (6) (2018) 5217–5230.

[33] Z. Wu, Q. Xu, J. Li, C. Fu, Q. Xuan, Y. Xiang, Passive indoor localization based on csi and naive bayes classification, IEEE Trans. Syst. Man Cybernet. Syst. 48 (9) (2017) 1566–1577.

[34] S. Shi, S. Sigg, Y. Ji, Probabilistic fingerprinting based passive device-free localization from channel state information, in: 2016 IEEE 83rd Vehicular Technology Conference (VTC Spring), IEEE, 2016, pp. 1–5.

[35] J. Xiao, K. Wu, Y. Yi, L.M. Ni, FIFS: fine-grained indoor fingerprinting system, in: 2012 21st International Conference on Computer Communications and Networks (ICCCN), IEEE, 2012, pp. 1–7.

[36] M.Z. Khan, A. Taha, W. Taylor, M.A. Imran, Q.H. Abbasi, Non-invasive localization using software-defined radios, IEEE Sensors J. 22 (9) (2022) 9018–9026.

[37] X. Wang, L. Gao, S. Mao, PhaseFi: phase fingerprinting for indoor localization with a deep learning approach, in: 2015 IEEE Global Communications Conference (GLOBECOM), IEEE, 2015, pp. 1–6.

[38] H. Chen, Y. Zhang, W. Li, X. Tao, P. Zhang, ConFi: Convolutional neural networks based indoor Wi-Fi localization using channel state information, IEEE Access 5 (2017) 18066–18074.

[39] X. Wang, X. Wang, S. Mao, ResLoc: deep residual sharing learning for indoor localization with CSI tensors, in: 2017 IEEE 28th Annual International Symposium on Personal, Indoor, and Mobile Radio Communications (PIMRC), IEEE, 2017, pp. 1–6.

[40] L. Batistić, M. Tomic, Overview of indoor positioning system technologies, in: 2018 41st International Convention on Information and Communication Technology, Electronics and Microelectronics (MIPRO), IEEE, 2018, pp. 0473–0478.

[41] S. Sadowski, P. Spachos, RSSI-based indoor localization with the Internet of Things, IEEE Access 6 (2018) 30149–30161.

[42] WiFi, 2022. (accessed 11.12.22) https://www.wi-fi.org/discover-wi-fi/specifications.

[43] J.P.G. Campeón, S. López, S.R. de Jesús Meleán, H. Moldovan, D.R. Parisi, P.I. Fierens, Indoor Positioning based on RSSI of WiFi signals: how accurate can it be? in: 2018 IEEE Biennial Congress of Argentina (ARGENCON), IEEE, 2018, pp. 1–8.

[44] M. Abbas, M. Elhamshary, H. Rizk, M. Torki, M. Youssef, WiDeep: WiFi-based accurate and robust indoor localization system using deep learning, in: 2019 IEEE International Conference on Pervasive Computing and Communications (PerCom), IEEE, 2019, pp. 1–10.

[45] J. Bai, Y. Sun, W. Meng, C. Li, Wi-Fi fingerprint-based indoor mobile user localization using deep learning, Wirel. Commun. Mob. Comput. 2021 (2021).

[46] RFID News Roundup, 2022. https://www.rfidjournal.com/tag/Internet-of-Things, (accessed 11.12.22).

[47] C. Li, L. Mo, D. Zhang, Review on UHF RFID localization methods, IEEE J. Radio Freq. Identif. (RFID) 3 (4) (2019) 205–215.

[48] L.M. Ni, Y. Liu, Y.C. Lau, A.P. Patil, LANDMARC: indoor location sensing using active RFID, in: Proceedings of the First IEEE International Conference on Pervasive Computing and Communications, 2003.(PerCom 2003), IEEE, 2003, pp. 407–415.

[49] C.-H. Huang, L.-H. Lee, C.C. Ho, L.-L. Wu, Z.-H. Lai, Real-time RFID indoor positioning system based on Kalman-filter drift removal and Heron-bilateration location estimation, IEEE Internet Things J. 64 (3) (2014) 728–739.

[50] S. Siachalou, S. Megalou, A. Tzitzis, E. Tsardoulias, A. Bletsas, J. Sahalos, T. Yioultsis, A.G. Dimitriou, Robotic inventorying and localization of rfid tags, exploiting phase-fingerprinting, in: 2019 IEEE International Conference on RFID Technology and Applications (RFID-TA), IEEE, 2019, pp. 362–367.

[51] S. Djosic, I. Stojanovic, M. Jovanovic, T. Nikolic, G.L. Djordjevic, Fingerprinting-assisted UWB-based localization technique for complex indoor environments, Expert Syst. Appl. 167 (2021) 114188.

[52] K. Yu, K. Wen, Y. Li, S. Zhang, K. Zhang, A novel NLOS mitigation algorithm for UWB localization in harsh indoor environments, IEEE Trans. Veh. Technol. 68 (1) (2018) 686–699.

[53] M. Werner, M. Kessel, C. Marouane, Indoor positioning using smartphone camera, in: 2011 International Conference on Indoor Positioning and Indoor Navigation, IEEE, 2011, pp. 1–6.

[54] W. Chen, W. Wang, Q. Li, Q. Chang, H. Hou, A crowd-sourcing indoor localization algorithm via optical camera on a smartphone assisted by Wi-Fi fingerprint RSSI, Sensors 16 (3) (2016) 410.

[55] Y. Xia, C. Xiu, D. Yang, Visual indoor positioning method using image database, in: 2018 Ubiquitous Positioning, Indoor Navigation and Location-Based Services (UPINLBS), IEEE, 2018, pp. 1–8.

[56] M. Zhao, M. Yan, T. Li, Vision-based positioning: related technologies, applications, and research challenges, in: 2018 IEEE 9th International Conference on Software Engineering and Service Science (ICSESS), IEEE, 2018, pp. 531–535.

[57] Bluetooth. https://www.bluetooth.com/bluetooth-resources/bluetooth-core-specification-v5-1-feature-overview/, (accessed 11.12.22).

[58] F.S. Danış, A.T. Cemgil, Model-based localization and tracking using Bluetooth low-energy beacons, Sensors 17 (11) (2017) 2484.

[59] S.G. Ndzukula, T.D. Ramotsoela, B.J. Silva, G.P. Hancke, A Bluetooth Low Energy based system for personnel tracking, in: IECON 2017-43rd Annual Conference of the IEEE Industrial Electronics Society, IEEE, 2017, pp. 8435–8440.

[60] Y. Yun, J. Lee, D. An, S. Kim, Y. Kim, Performance comparison of indoor positioning schemes exploiting Wi-Fi APS and BLE beacons, in: 2018 5th NAFOSTED Conference on Information and Computer Science (NICS), IEEE, 2018, pp. 124–127.

[61] S. Subedi, S.-S. Hwang, J.-Y. Pyun, Hybrid wireless indoor positioning system combining BLE beacons and Wi-Fi APs, in: 2020 International Conference on Information and Communication Technology Convergence (ICTC), IEEE, 2020, pp. 36–41.

[62] W. Song, J. Liao, J. Han, A real-time human posture recognition system using Internet of Things (IoT) based on LoRa wireless network, in: Advances in Computer Science and Ubiquitous Computing, Springer, 2021, pp. 379–385.

[63] LoRa-Alliance. https://lora-alliance.org/, (accessed 12.12.22).

[64] K.-H. Lam, C.-C. Cheung, W.-C. Lee, RSSI-based LoRa localization systems for large-scale indoor and outdoor environments, IEEE Trans. Veh. Technol. 68 (12) (2019) 11778–11791.

[65] M.T. Hoang, B. Yuen, X. Dong, T. Lu, R. Westendorp, K. Reddy, Recurrent neural networks for accurate RSSI indoor localization, IEEE Internet Things J. 6 (6) (2019) 10639–10651.

[66] C. Jiang, J. Shen, S. Chen, Y. Chen, D. Liu, Y. Bo, UWB NLOS/LOS classification using deep learning method, IEEE Commun. Lett. 24 (10) (2020) 2226–2230.

[67] T. Koike-Akino, P. Wang, M. Pajovic, H. Sun, P.V. Orlik, Fingerprinting-based indoor localization with commercial MMWave WiFi: a deep learning approach, IEEE Access 8 (2020) 84879–84892.

[68] P. Roy, C. Chowdhury, Designing an ensemble of classifiers for smartphone-based indoor localization irrespective of device configuration, Multimed. Tools Appl. 80 (13) (2021) 20501–20525.

[69] L. Zhang, Z. Chen, W. Cui, B. Li, C. Chen, Z. Cao, K. Gao, WiFi-based indoor robot positioning using deep fuzzy forests, IEEE Internet Things J. 7 (11) (2020) 10773–10781.

[70] W. Liu, H. Chen, Z. Deng, X. Zheng, X. Fu, Q. Cheng, LC-DNN: local connection based deep neural network for indoor localization with CSI, IEEE Access 8 (2020) 108720–108730.

[71] A. Christy Jeba Malar, M. Deva Priya, F. Femila, S.S. Peter, V. Ravi, Wi-Fi fingerprint localization based on multi-output least square support vector regression, in: Intelligent Systems, Springer, 2021, pp. 561–572.

[72] E. Schmidt, D. Akopian, Indoor positioning system using WLAN channel estimates as fingerprints for mobile devices, in: Mobile Devices and Multimedia: Enabling Technologies, Algorithms, and Applications 2015, vol. 9411, SPIE, 2015, pp. 239–247.

[73] L. Yin, T. Jiang, Z. Deng, Z. Wang, Improved fingerprint localization algorithm based on channel state information, in: 2019 IEEE 1st International Conference on Civil Aviation Safety and Information Technology (ICCASIT), IEEE, 2019, pp. 171–175.

[74] N.A.M. Maung, B.Y. Lwi, S. Thida, An enhanced RSS fingerprinting-based wireless indoor positioning using random forest classifier, in: 2020 International Conference on Advanced Information Technologies (ICAIT), IEEE, 2020, pp. 59–63.

[75] J. Liu, N. Liu, Z. Pan, X. You, AutLoc: deep autoencoder for indoor localization with RSS fingerprinting, in: 2018 10th International Conference on Wireless Communications and Signal Processing (WCSP), IEEE, 2018, pp. 1–6.

[76] Y.J. Wang, Y. Wang, Y. Zhang, Indoor positioning algorithm for WLAN based on KFCM-LMC-LSSVM, Acta Metrol. Sin. 39 (4) (2018) 554–558.

[77] A. Basiri, E.S. Lohan, T. Moore, A. Winstanley, P. Peltola, C. Hill, P. Amirian, P.F. e Silva, Indoor location based services challenges, requirements and usability of current solutions, Comput. Sci. Rev. 24 (2017) 1–12.

[78] Estimote, 2022. https://https://estimote.com/, (accessed 12.12.22).

[79] Infsoft, 2022. https://www.infsoft.com/, (accessed 12.12.22).

[80] Inpixion, 2022. https://www.inpixon.com/, (accessed 12.12.22).

[81] MOKOSmart, 2022. https://www.mokosmart.com/documentation/, (accessed 12.12.22).

[82] Zebra, 2022. https://www.zebra.com/, (accessed 12.12.22).

About the authors

Muhammad Zakir Khan, an alumnus of CUSIT, earned his BCS (Hons) in computer science in 2016, followed by an MS (SE) in Software Engineering in 2019. Currently engaged in a Ph.D. program at the University of Glasgow's James Watt School of Engineering, Zakir is passionately delving into the realm of Electronic and Electrical Engineering. His profound research pursuits encompass the dynamic domains of machine learning, data mining, and innovative clustering approaches. His expertise extends to exploring contactless indoor localization, seeking groundbreaking advancements in this area.

Muhammad Farooq is currently pursuing a Ph.D. in Electronics & Electrical Engineering with the Communication, Sensing, and Imaging group at the James Watt School of Engineering, University of Glasgow, UK. His research focuses on noninvasive human activity recognition and human vitals detection for contactless healthcare systems. Farooq holds a master's degree in electrical engineering, specializing in systems, control, and robotics from Lahore University of Management Sciences (LUMS), Pakistan. Beyond his academic achievements, Farooq has garnered extensive professional experience in the engineering field. He worked as a Project Lead and Research Associate, contributing significantly to the development of noninvasive healthcare systems. Additionally, his expertise extended to the field of industrial process automation during his tenure as an Application & Design Engineer with the Endress+Hauser Group. His diverse expertise and dedication to innovative technologies have established him as a prominent figure in both academic and industrial realms.

Ahmad Taha is a lecturer in Autonomous Systems and Connectivity at the James Watt School of Engineering, University of Glasgow, with over a decade of experience across esteemed higher education institutions spanning Egypt, the UK, and China. His research journey centres on cyber-physical systems for energy management and digital healthcare. Dr. Taha has authored/ coauthored over 35 publications showcased in reputable venues and has contributed as a Principal Investigator/Co-Investigator to research grants exceeding £350k in value. Dr. Taha has earned recognition from the Royal Academy of Engineering through the Global Talent scheme, he is an inaugural member of the UK Young Academy and an academic advisor to the Commonwealth Scholarship Commission. Dr. Taha holds the title of Fellow of Advanced Higher Education (FHEA) and is among the pioneering UKCGE recognised Associate Supervisors.

Adnan Qayyum is a Ph.D. candidate in Computer Science at the Information Technology University (ITU), Lahore, Pakistan. His research interests include inverse medical imaging problems, healthcare, and secure, robust, and trustworthy ML. He received the bachelor's degree in Electrical (Computer) Engineering from COMSATS Institute of Information Technology, Wah, Pakistan, in 2014 and M.S. degree in Computer Engineering (Signal and Image Processing) from the University of Engineering and Technology, Taxila, Pakistan, in 2016.

Fehaid Alqahtani received the B.Sc. degree in computer science from King Khalid University, Saudi Arabia, in 2008, the M.Sc. degree in IT project management from Teesside University, UK, in 2014, and the Ph.D. degree in Artificial Intelligence (AI) from the University of the West of Scotland, UK, in 2020. He is currently an Assistant Professor with the Department of Computer Science, King Fahad Naval Academy, Ministry of Defense, Saudi Arabia. He has authored or coauthored several research papers, including leading international journals and peer-reviewed international conference proceedings. His research interests include affective computing and machine learning. He is also a Leading Team Member with the AI Society research committee, Member in association of technology systems and member of the social support association (Moazarah) specializes in providing technical support and consultations in Kingdom of Saudi Arabia.

Adnan Nadeem Al Hassan received his Ph.D. degree from the Institute for Communication Systems, the University of Surrey in the U.K. in 2011. He has been serving as an associate professor in the Faculty of Computer and Information Systems (FCIS), at the Islamic University of Madinah since 2016. He has a registered patent & published more than 90 research papers in international journals and conferences with a cumulative impact factor of 86.05. He has more than 2059 citations, and his Researchgate score is above 96% of RG members. He has completed 8 funded research projects as PI & CO-PI. He received several awards including Outstanding Researcher, Best Paper award, and Best Academic Advisor award. He is a member of IEEE and IEEE Communication Society & various scientific societies and serves as a review editor in international journals. He has mainly worked with IoT, AI, and imaging technologies applied to Healthcare, Smart Cities, and Agriculture. Also work on Security, Intrusion Detection, and Response Systems for wireless networks, BlockChain Technology, and assistive living applications.

Kamran Arshad is a senior member of the Institute of Electrical and Electronics Engineers (IEEE), currently holds the distinguished position of Dean of Research and Graduate Studies and Professor of Electrical Engineering at Ajman University in the United Arab Emirates. Dr. Arshad has played a pivotal role in leading numerous locally and internationally funded research projects that encompass the fields of cognitive radio, LTE/LTE-Advanced, 5G, optimization, and cognitive machine-to-machine communications. He has made significant contributions to several European and international large-scale research projects and has over 150 technical peer-reviewed articles published in esteemed journals and international conferences. He has been the recipient of three Best Paper Awards, one Best Research and Development Track Award, and has chaired technical sessions in several leading international conferences. Dr. Arshad holds the position of Associate Editor of the Journal on Wireless Communications and Networking (EURASIP).

Khaled Assaleh is currently the Vice Chancellor for Academic Affairs and a Professor of Electrical Engineering at Ajman University at Ajman University. From 2002 through 2017, he was with the American University of Sharjah (AUS) as a Professor of Electrical Engineering and Vice Provost for Research and Graduate Studies. Prior to joining AUS, Khaled had a 9-year R&D career in the Telecom Industry in the USA with Rutgers, Motorola and Rockwell/Skyworks. He earned a Ph.D. in Electrical Engineering from Rutgers, the State University of New Jersey in 1993; a Master's degree in Electronic Engineering from Monmouth University, New Jersey in 1990; and a B. Sc. in Electrical Engineering from the University of Jordan in 1988. He holds 11 US patents and has published over 150 articles in signal/image processing and pattern recognition and their applications. He has served on organization committees of several international conferences including ICIP, ISSPA, ICCSPA, MECBME and ISMA. He has also served as a guest editor for several special issues of journals. Dr. Assaleh is a senior member of the IEEE. His research interests include biosignal processing, biometrics, speech and image processing, and machine learning.

Shuja Ansari is a Lecturer in Autonomous Systems and Connectivity at the James Watt School of Engineering, University of Glasgow. He is the Glasgow principal investigator for the UK EME Hub, a senior member IEEE and a chartered engineer with a strong background in telecommunications and its applications. He has a research grant portfolio of over £3m as principal and coprincipal investigator and has published over 60 high quality research articles. He is the 5G and IoT use case lead for the Wave-1 Urban 5G project funded by the Scotland 5G Centre and project manager for Glasgow COMPORAN, Glasgow ONSIDE and Glasgow BEACH projects funded by DSIT. He is also the deputy director of impact, leading the PGR and Postdoc engagement at the School of Engineering. His research interests include resilient and robust wireless communications, digital twins, systems integration, terrestrial/airborne mobile networks, security and privacy in communications and Internet of Things (IoT) applications.

Muhammad Usman is a Lecturer and Programme Leader in the Department of Electrical and Electronic Engineering at Glasgow Caledonian University (GCU). Before Joining GCU, he was a Research Associate at the University of Glasgow, UK, in the Engineering and Physical Sciences Research Council (EPSRC) funded COG-MHEAR programme grant. This transformative research aims to redesign current hearing aids and assistive technology, where his role was to integrate end-user context in visually assisted hearing-aid design in a privacy-preserving manner.

Muhammad Usman is endorsed by the Royal Academy of Engineering as Global Talent. He holds a Ph.D. degree (cum laude) in information and communication technologies from the University of Trento, Italy. He has as exceptional research output record in the field of 5G and beyond 5G cellular systems, radio frequency (RF) sensing, network security, Internet-of-Things (IoT) and developing intelligent systems for healthcare sector. He is a senior member of IEEE. His research interests include RF sensing, wireless communication, software-defined networks, cyber security and assistive technologies for intelligent healthcare systems.

Muhammad Ali Imran (senior member, IEEE) is currently the Dean of the University of Glasgow UESTC, Glasgow, UK, and a Professor of wireless communication systems. He also heads the Communications, Sensing and Imaging Research Group with the University of Glasgow, and is the Director of Glasgow-UESTC Centre for Educational Development and Innovation. He is an Affiliate Professor with The University of Oklahoma, Norman, OK, the USA, and a Visiting Professor with 5G Innovation Centre, University of Surrey, Guildford, UK. He has more than 20 years of combined academic and industry experience with several leading roles in multimillion pounds funded projects. He has filed 15 patents, has authored, or coauthored more than 400 journal and conference publications, edited seven books and authored more than 30 book chapters, successfully supervised more than 40 postgraduate students at Doctoral level. His research interests include self-organised networks, wireless networked control systems, Internet of Things, and wireless sensor systems. He has been a consultant to international projects and local companies in self-organised networks. He is a Fellow of IET, and Senior Fellow of HEA.

Qammer H. Abbasi (senior member, IEEE) is currently a Reader with the James Watt School of Engineering, University of Glasgow, Glasgow, UK, and the Deputy Head for Communication Sensing and Imaging Group. He has authored or co-authored more than 350 leading international technical journals and peer-reviewed conference papers and ten books. He was the recipient of several recognitions for his research, including URSI 2019 Young Scientist Awards, UK exceptional talent endorsement by Royal Academy of Engineering, Sensor 2021 Young Scientist Award, National talent pool award by Pakistan, International Young Scientist Award by NSFC China, National interest waiver by USA and eight best paper awards. He is a Committee Member of IEEE APS Young professional, Subcommittee Chair of IEEE YP Ambassador program, IEEE 1906.1.1 standard on nano communication, IEEE APS/SC WG P145, IET Antenna & Propagation, and healthcare network. He is also a member of IET and a Fellow of RET and RSA.

CHAPTER FOUR

Smart indoor air quality monitoring for enhanced living environments and ambient assisted living

Jagriti Saini[a] , Maitreyee Dutta[b] , and Gonçalo Marques[c]
[a]AMIE, IEEE Member, Mandi, Himachal Pradesh, India
[b]National Institute of Technical Teacher's Training and Research, Chandigarh, India
[c]Polytechnic Institute of Coimbra, Technology and Management School of Oliveira do Hospital, Oliveira do Hospital, Portugal

Contents

Abstract

Indoor air quality has been a major concern for public health and well-being. People who spend 80–90% of their routine time indoors are likely to experience serious complications due to repeated exposure to degraded air quality. Patients with disability and existing health concerns usually interact more with the indoor environment. Therefore, the risk of rising concentration of air pollutants in the building premises in urban and rural areas brings a considerable threat to people with assisted living. This chapter focuses on the importance of smart indoor air quality monitoring for enhanced living environments and ambient assisted living. It throws light on the efforts made by past researchers, potential gaps in the literature, future scopes, and technical recommendations to deal with the problem domain. Furthermore, this chapter also introduces the latest technologies, methods, and approaches to support enhanced living environments and ambient assisted living.

Advances in Computers, Volume 133
ISSN 0065-2458
https://doi.org/10.1016/bs.adcom.2023.10.008

1. Introduction

Degraded air quality has been a matter of concern worldwide, especially among public health professionals, and policymakers. On average, human beings spend almost 90% of their routine time indoors; therefore, thermal comfort and indoor air quality are the key factors affecting occupant health, comfort, and performance [1]. Statistics reveal that in developed countries, indoor air pollution (IAP) has more serious health impacts in comparison to outdoor air pollution [2]. In the year 2010, IAP from solid fuels lead to 4.5% global daily adjusted life years (DALY) along with approximately 3.5 million deaths [3]. The indoor environmental conditions are influenced by a variety of several ambient parameters such as air movement, surrounding surface temperature, indoor air temperature, relative humidity, rate of air exchange, carbon dioxide, and pressure. Substances that are often present in the indoor environment include particulate matter (PM_{10}, $PM_{2.5}$, and $PM_{0.1}$ based on particle aerodiameter), sulfur dioxide, terpenes, ketones, aldehydes, volatile organic compounds (VOCs), carbon monoxide (CO), nitrogen dioxide (NO_2), ozone (O_3), asbestos, radon, and bio contaminants such as fungi, bacteria, molds, viruses, animal dander and allergens [4]. In middle and low-income countries, the majority of populations prefer using solid fuels for cooking and heating needs which is the main cause of enhanced mortality and morbidity rates [5]. Children, women, the elderly, and assisted living patients who spend more time indoors are the people with the highest health risk due to degraded air quality.

The primary sources of IAP include building materials, combustion, and bioaerosols. The combustion products are more common in developing countries whereas pesticides, asbestos, radon, VOCs, heavy metals, and environmental tobacco smoke are more prevalent in developed countries [6]. In India, approximately 0.2 billion people are still using biomass fuel for routine cooking needs; out of which 8.9% use cow dung cake; 2.9% use kerosene; 49% firewood; 28.6% liquefied petroleum gas; 0.4% biogas; 0.1% electricity; 1.5% charcoal, lignite or coal; and remaining 5% use any other means [3]. The biological and chemical reactivity of pollutants in the human body can lead to huge damage to the molecular and cellular mechanisms. The serious health effects of the toxicants released by burning solid fuels and repeated use of chemical products include chronic obstructive respiratory disease (COPD), acute respiratory infections (ARI), cancer of the larynx and nasopharynx, lung cancer, perinatal conditions, cardiovascular disease, tuberculosis, blindness, cataract and low birth weight [7]. Air pollutants can also modulate human immune responses due to dangerous respiratory viral infections.

In order to deal with the increasing threats due to IAP, researchers and policymakers across the world are making efforts to utilize the latest technologies for improved management of IAQ. Advanced information and communication technologies, sensor networks, and artificial intelligence-based software systems have opened up new doors for enhanced environmental monitoring and management [8]. The novel smart buildings can be designed to reduce the impact of IAP with automated ventilation systems and early prediction systems. The open-source software, microcontrollers, sensors, and data visualization approaches enable 24-h monitoring and enhanced control of indoor living environments.

This chapter throws light on the latest technologies, methods, and approaches available with smart building management systems for enhanced living environments and ambient assisted living. Section 2 below describes the importance of measuring IAQ for enhanced living environments. Section 3 further shows the potential of hardware architectures and sensor systems for IAQ monitoring. Furthermore, various technology-based methods and approaches for IAQ management in different types of buildings are discussed in Section 4. Ultimately, Section 5 presents the conclusions.

2. Indoor air quality for enhanced living environments

The respiratory health problems associated with IAP in the general population have been documented in existing literature. But less attention is paid to elderly people, assisted living patients living at nursing homes, care centers, and elderly centers. Due to increased life expectancy and reduced fertility rates. It is observed that the elderly, especially those above the age group of 80 years are highly affected by health impairments caused by IAP in poorly ventilated buildings [1]. One research study conducted by [9] focused on 10 different elderly care centers located in different parts of Portugal show increased concentrations of hazardous substances such as PM_{10}, O_3, VOCs, and CO_2 due to poor ventilation arrangements in the rooms. Different types of hazardous pollutants that are commonly found in assisted living premises are discussed in the below subsection.

2.1 Types of pollutants

There are generally three factors that play an critical role in deciding IAP levels in any type of building. It includes gaseous pollutants, particulate matter or dust particles, and thermal comfort parameters. The most commonly found compounds in nursing homes and elderly care centers include carbon dioxide, nitrogen oxides, carbon monoxide, ozone, and sulfur dioxides. For elderly care centers and assisted living homes, CO_2 assessment is of utmost importance since this is the most hazardous pollutant for susceptible populations. ASHRAE guidelines establish a CO_2 concentration of less than 1000 ppm for indoor spaces; however, in the context of the recent COVID-19 pandemic, 700 ppm was declared as the safe level for indoor settings [10]. Furthermore, the investigations conducted by existing researchers show that indoor CO_2 levels in assisted living environments are often greater than safe levels. The readings were influenced by a variety of conditions including building materials, surrounding conditions, indoor activities, and infiltration rate [9]. The authors of [11] reported the estimated concentration of CO_2 between 561 and 850 ppm in the summer season whereas it was between 792 and 1881 ppm during winter. The greater CO_2 levels are further linked to the increased risk of breathlessness among elderly people at nursing homes [12].

The main source of carbon monoxide production is the incomplete combustion of biomass fuels such as wood, coal, petrol, and natural gas. WHO recommendations state four different mean periods for CO concentrations including $4 \, mg/m^3$ (24h), $10 \, mg/m^3$ (8h), $35 \, mg/m^3$ (1 h), and $100 \, mg/m^3$ (15 min) [13]. Several studies conducted by existing researchers show that CO mean concentrations are often greater than WHO recommendations in the bedroom areas of elderly care facilities [4,11]. The main sources of CO pollution in such areas were cooking activities, tobacco smoke, and outdoor elements such as traffic. In urban areas where most of the elderly care facilities are near high-traffic regions with poor ventilation arrangements, the CO levels were reported to be quite high throughout the year. Repeated exposure to higher CO levels is further linked to COPD in elderly patients [12]. NO_2 is another hazardous gaseous pollutant that is commonly found in urban areas and megacities. WHO guidelines advise regulation of NO_2 on three mean periods as $10 \, \mu g/m^3$ (annual mean), $25 \, \mu g/m^3$ (24 h mean), and $200 \, \mu g/m^3$ (1 h mean). The increased concentration of this pollutant in elderly care facilities often leads to breathlessness and cough. The presence of VOCs in the indoor environment can further cause several inhalation-related problems. The major sources of VOCs are fireplaces, candles, cooking activities, cleaning products, furniture, and many other building materials [1]. The reaction of VOCs with other oxidants such as NO_2 further leads to the generation of indoor ozone which is another dangerous pollutant for the building environments [13].

Particulate matter is another critical parameter that is required to be assessed for both indoor and ambient environments. Nevertheless, the exposure assessment for PM leads to extra difficulty due to the variable range of particle sizes and unique chemical compositions [14]. Airborne particles with a particle diameter somewhere around 2.5 μm are called $PM_{2.5}$ pollutants. The smaller size of these particles allows easy penetration into respiratory systems while leading to some adverse impacts on human health. $PM_{2.5}$ levels can be influenced by several factors such as construction characteristics, insulation or flooring arrangements, building pathologies, ventilation arrangements, and seasonal variations as well [15]. The existing studies claim higher $PM_{2.5}$ concentrations in the elderly centers in comparison to the compliance provided by WHO guidelines with a mean limit of 15 μg/m^3 for 24 h period [13–15]. Similarly, particles with a size close to the 10 μm range (PM_{10}) are known as coarse particles and WHO established mean limits for these particles is 50–45 μg/m^3 for the 8 h period [13]. Higher concentrations of PM_{10} levels are generally reported in indoor environments with increased occupancy rates. The presence of these particles in assisted care facilities leads to several respiratory health problems among people. The thermal comfort parameters and biological contaminants pose another level of risk to older people and disabled patients. Along with critical levels of temperature and humidity levels, the presence of a variety of allergens, viruses, fungi, bacteria, and toxins leads to critical health outcomes [16]. The impact further varies with the type of ventilation arrangements, air exchange rates, and living conditions in the target building environment. Apparently, chronic disease turn out to be more prevalent as people get older due to reduced immunity levels. In such scenarios, the repeated exposure to hazardous pollutants can worsen the conditions with more common cases of cancer, cardiopulmonary diseases and respiratory illnesses [17].

Global stats reveal that there are almost 700 million people across the world with an age group greater than 65. The percentage is likely to increase up to 29.4% by the end of 2050 [18]. As per the stats obtained from the United Nations in the year 2012, 22% of the population in Europe was from 60 years of age group and the projections for 2050 are 34%. Furthermore, Portugal is rated as the 8th oldest nation in the world with 6th rank in Europe as the country has around 23% population above 60 years of age group [9]. The aged populations due to reduced mobility spend more time in indoor environments, especially in assisted living premises, nursing homes, or elderly care centers. Therefore, they are always at high risk of being exposed to the IAP.

Ensuring adequate IAP levels with enhanced thermal comfort is necessary to prevent deaths among vulnerable population groups. Thermal comfort is a subjective concept because this sensory experience may vary from person to person. It is complex to assess thermal comfort levels due to the great interdependence of variables [1]. The main characteristics that influence this factor include clothing insulation and metabolic rates of individuals. Usually, residential spaces do not require special devices for controlling humidity levels since it varies with human occupation and air exchange rates. But high humidity levels can often lead to the development of various microorganisms, especially fungi, which further contributes to various types of pneumonia cases, skin illnesses, and respiratory problems. Several characteristics of the built environment play an essential role in deciding IAP exposure at assisted living centers. The existing operational systems, construction materials, building designs, routine maintenance procedures, occupant activities, the density of occupants, and the products that are used on a regular basis are likely to affect the concentration of air pollutants in indoor environments. Furthermore, several thermal comfort parameters such as temperature, humidity, and wind conditions can further intensify the presence of hazardous pollutants. Field experts always recommend ensuring adequate ventilation arrangements for enhanced living environments to improve the life expectancy of people with mobility and physical disabilities. The design aspects of a building can play an essential role in this process. It is necessary to pay attention to the building openings, window sizing, orientation, and vents providing natural sunlight along with circulation of fresh air [19]. The presence of nearby industrial units, high-traffic areas as well as green spaces such as flowers, shrubs, gardens, and green walls has been also studied by existing researchers. It is observed that assisted living homes that are close to green spaces are likely to have better filtration for air pollutants in comparison to buildings that are close to the main roads and industrial units [20]. It is necessary to establish a correlation between critical IAQ parameters and building characteristics. Several properties of construction materials and components have to be studied along with the type of window sealants, heating systems, ventilation types, and condensations. Existing studies show that elderly centers and nursing homes are not projected to adapt IAQ requirements for buildings.

Ambient assisted living is an area of interest for researchers across the world these days, especially those who are making efforts to enhance the well-being of older adults and people with disabilities. The idea is to develop an ecosystem by leveraging the benefits of the latest technologies to design reliable monitoring and control mechanisms with ubiquitous computing. The new algorithms, platforms, and innovative applications can promise a reliable autonomous living experience. Furthermore, the effective integration of software and hardware can help to deal with the healthcare consequences influenced by the degraded air quality in indoor environments. It is crucial to monitor environmental parameters in the buildings so that relevant control actions can be taken on time. The real-time data-sharing capabilities of wireless communication technologies such as ZigBee, Ethernet, Wi-Fi, and 5G can assist with quick updates about critical parameter concentrations so that possible interventions can be taken to enhance health and productivity levels of the individuals in the enhanced living environments [21]. These ideas can be also utilized to lead smart city projects while supporting the health and well-being of all age groups. Since active environmental monitoring and control can improve sustainability, it is possible to make better policies and design decisions for building architectures. The sections below will highlight the potential of the latest technologies at multiple levels to enhance the quality of environments at assisted living facilities.

3. Indoor air quality monitoring: Hardware architectures and sensors

In order to control and manage environmental quality in the assisted living premises, it is mandatory to understand the characteristics of the pollutants in the target buildings. There is a great need of deploying cost-effective and reliable sensor units that can measure field data on a regular basis while guiding relevant actions for the control operations. Usually, the air pollution concentrations at commercial buildings and government offices are measured using professional monitoring stations that promise the utmost accuracy in measurements. But such systems are highly expensive and cannot be installed in all types of buildings in smart city projects. Furthermore, operations of such high-end sensors are also very costly with periodic maintenance requirements from trained engineers. Studies also reveal that pollutant concentrations and chemical compositions also keep on varying from one building to another depending upon structure, surrounding environments, and living conditions. A common sensor setup cannot be used in all regions equally. Rather, the upcoming researchers need to access independent scenarios for the building environments in rural and urban areas and design custom hardware architectures to meet specific requirements. The requirements are very specific in assisted living environments where improper environmental conditions can pose serious threats to the majority of the world's population which includes elderly people.

The low-cost IAQ monitoring solutions for smart buildings consist of a range of sensing units. Along with the atmospheric sensors that are responsible for measuring pollutants, there are a few additional hardware and software such as processing units, power sources, networking interfaces, and local data storage. Depending upon the number and type of pollutants that are measured, it may require a set of sensors with different specifications. Experts always recommend checking manufacturer datasets before buying any sensor for IAQ monitoring so that the operating conditions and measurement ranges can be verified in advance. At the same time, datasheets can help researchers know about underlying sensor technology, working principles, key advantages, calibration requirements, and shortcomings of a sensing unit with respect to the target field environment. Before talking about these details, it is important to discuss the variety of atmospheric sensors that can be utilized to measure IAQ levels in enhanced living environments.

3.1 Sensors for IAP measurement

Characteristics of the sensing units and overall design of the hardware monitoring system play a critical role in IAQ assessment. Therefore, all units must be carefully selected to meet the field requirements. Depending upon the type of pollutants required to be measured from the assisted living environments, researchers can go ahead with several different sensor units. In general, gaseous pollutants can be measured by four different sensing technologies—photoionization detector (PID) sensors, non-dispersive infra-red (NDIR) sensors, electrochemical (EC) sensors, and metal oxide semicon-ductor (MOS) sensors. MOS sensors can be used to measure concentrations of several gaseous pollutants including O_3, NO_x, CO_2, NO_2, CO, and non-methane hydrocarbon. MOS sensors are constituted of a semiconducting metal oxide sensing element and a heating element. The heater on the sensor raises the sensing element temperature up to 300–500 °C which leads to chemical reactions on the surface that helps in detecting relevant gases [22]. One external circuit can be used to measure the gas levels with respect to changes in the electrical conductivity of the sensing element. MOS sen-sors are widely recommended due to their short response time and high sen-sitivity. At the same time, they show higher resistance against extreme weather conditions which makes them suitable for long-term monitoring applications. However, the main problem with these sensors is their repeated calibration requirement which often leads to low-quality measurements. Furthermore, the high power requirements of these sensors make them suitable to deploy in areas with reliable and permanent power arrangements since one cannot rely on batteries. Although recent studies show the utili-zation of new materials for MOS sensor design that promises better sensitiv-ity even when humidity levels increase, such materials are not yet available for bulk production [23].

EC sensors on the other side are widely recommended to monitor O_3, NO_2, NO, CO, and SO_2. They work by using oxidation–reduction reactions while using electrodes segregated by an effective electrolyte sub-stance. When the working electrode comes in contact with the ambient air and electrolyte, the reaction leads to the production of an electrical current that can be ultimately measured on the sensor terminals. These low-cost sensors are known to have higher specificity and sensitivity levels while being less affected by humidity and temperature variations in the field environment. But the main disadvantage of EC sensors is that the effective-ness of chemical reactions is highly influenced by the operating conditions. Moreover, the operating range of these sensors is also slightly lower than the MOS sensors. In terms of sensor calibration, both MOS and EC sensors require re-validation time and again. The average lifetime of EC sensors is limited to 2–5 years; however, MOS sensors can perform well for up to approximately 10 years. NDIR sensors work with an infrared light source, an optical filter, and an atmospheric sampling chamber. Whenever gas enters the chamber, the light from the infrared source travels through the channel, and depending upon the type of gas, some frequencies get absorbed. The optical filter further receives the light signal and converts it into an equivalent electrical current. These sensors are known to have a long lifespan and they need very little maintenance with little power requirement. However, their detection limits are generally high; therefore, cannot be applied in areas with small pollutant concentrations.

PID sensors work through the illuminating compounds while utilizing UV photons of considerable energy. This process leads to the ionization of compounds when they absorb UV photons and the resulting electrical current is captured by a detector within the sensor. These detectors are very sensitive and they offer short response time. With small size and weight, they also require lesser energy for operation. However, the major disadvantage of these sensors is the drift that triggers the requirement of re-calibration almost once every month. Generally, the choice of gas sensors depends upon the requirement of the target field where the hardware system has to be deployed. Existing studies reveal that EC and MOS sensors can be utilized for large-scale deployments due to their cheap availability; however, NDIR and PID sensors are better for sparser deployments. In order to measure gas concentrations in Ambient Assisted Living environments, it is important to be careful about the sensor calibration requirements as well.

In order to measure concentrations of PMs in nursing homes and elderly care centers, it is important to use sensors that are tailored to detect dust particles of specific sizes. However, many sensors are capable to measure particles of different size ranges. There are generally two popular technologies that are being used for the past several years to design PM sensors: Light Scattering Particle Sensors (LSPs) and Diffusion Size Classifiers (Discs). The LSPs work by using the light scattering principle to measure particle density; however, Discs work by charging the air in the sensor cavity to estimate the total density of the particle through an electric charge. These sensors apply a variety of filtration operations on charged air to ensure accurate measurements. Several laboratory-grade PM sensors are based on the LSP principle; however, many of them use some additional components to enhance measurement accuracy. For instance, condensation particle counters (CPCs) make use of water vapors or alcohol to alter the physical properties of the dust particles passing through sensors; however, optical particle counters (OPCs) are high-end variants of LSPs with enhanced accuracy levels ([24], p. 2). In a similar manner, the beta attenuation monitor (BAM) makes use of beta radiation absorption to estimate particle density;

whereas Tapered Element Oscillating Microbalance (TEOM) units use oscillation frequency changes within a vibrating glass tube to measure density. But both these sensors are very expensive and cannot be used for large-scale deployment. The Disc sensors are known for their high sensitivity and low power consumption. They are also available at low cost when manufactured in bulk amounts. However, the upfront cost for manufacturing is quite high as they need special facilities with clean rooms. On the other hand, LSPs-based sensors are available at low cost but they often fail to detect small-sized dust particles. The sensor accuracy is also affected by changes in environmental variables such as relative humidity and temperature. Weather has a great impact on particle concentration in the air, as a result, it may affect sensor measurements as well. This is the main reason why most PM sensors need adequate information about weather conditions, to ensure desired level of accuracy. The PM concentration is also gets affected by human activity in the target field environment. Other than this, researchers till now have used a variety of thermal comfort sensors that can measure temperature and relative humidity. Some of the most commonly used sensors are Negative Temperature Coefficient (NTC) type which includes DHT11, DHT22, and SHT11, SHT22. These sensors are factory calibrated; therefore, easy to interface with the microcontrollers.

The field measurements of low-cost environmental monitoring sensors can be recognized as time series data that consists of readings for different pollutants. In order to understand the impact of IAQ on human health in assisted living environments, it is necessary to analyze the impact of pollutant cross-sensitivity. Studies reveal that particulate matter and gaseous pollutants are the most prominent variables for measuring air quality indexes; whereas relative humidity and temperature are required to be considered to analyze the level of comfort for the building occupants [25,26]. At several locations, wind speed can be also considered as an influential variable. However, it is difficult to measure wind speed with low-cost sensors due to the requirement of unobstructed air intake, which is not possible on indoor premises. In general, calibration is the biggest concern while using sensors for IAQ measurement in enhanced living environments. Experts advise using calibrated sensors to access high-quality measurements. However, low-cost sensors are not always available in factory-calibrated form. In such situations, it is necessary to consider reference measurements from a high-quality source to work as a ground truth. Keeping low-cost sensors close to the reference station can help to establish calibration relationships more accurately. Furthermore, the temporal resolution of field measurements also influences the sensor calibration process. Sensor resolution is generally governed by sampling frequency which usually varies among different sensor technologies.

3.2 Hardware architectures for IAP measurement

IAP monitoring in assisted living environments requires designing a reliable, cost-effective, and sustainable hardware system. Advanced technologies such as the Internet of Things, wireless communication, and artificial intelligence can help to build managers and create comfortable solutions for enhancing the overall well-being of occupants. The new technologies make it easier to design intelligent monitoring solutions for regular building environments. Several researchers in the past have utilized wireless sensor network-based hardware designs with Zigbee and Bluetooth technologies for data transfer. There are several open-source software platforms that can be used for data collection from the field environment. On one hand, it is possible to use microcontrollers such as Arduino Uno, Raspberry Pi, and AT mega processors for data acquisition. On the other hand, mobile computing technologies can ensure easy access to field data from any location, at any time. Researchers these days are working on multi-disciplinary approaches to building reliable architectures with enhanced connectivity and design flexibility. In combination with WSN networks, researchers also preferred designing Android mobile applications and web portals to meet data storage as well as real-time monitoring requirements. The gateway units are added to the hardware to collect data from various sensor nodes and this information can be further utilized to analyze the conditions of assisted living environment. The WSN-based systems help people measure several different parameters from the target environment in real time. The researchers [27] measured several essential IAQ parameters such as temperature, dust particle concentrations, relative humidity, and acoustic levels with the help of GPY21010AU0F dust sensor, 808H5V5 humidity sensor, MCP9700A temperature sensor, and POM–2735P–R audio sensor. They used a ZigBee radio interface to transfer field data to a web data management platform. This WSN-based system was capable enough to monitor IAQ data at a large scale and the gathered information can be utilized to optimize air quality in the closed premises. Piatra et al. [28] proposed an IAQ monitoring system to collect random samples from the building environment to deal with the symptoms of sick building syndrome. The authors in this study used the Bee module, along with microsensors and Arduino to monitor real-time IAQ data including humidity, temperature, luminosity, and carbon monoxide. The sensors used for designing hardware systems by these researchers were SHT10, MQ7, T6615 CO_2 sensor, and LDR 5 mm sensor.

The authors [29] worked on the design of a WSN-based sensor network to monitor IAQ parameters in an office environment. The authors preferred using Sensation SCD 30 which includes a Nondispersive Infrared detection system to ensure accurate measurement of CO_2 gas. They also integrated one Sensation temperature and humidity sensor into the field unit. The data communication from the field environment was made possible with Bee pro modules The researchers preferred using MBED LPC 1768 chip for data acquisition and processing from sensors. Similarly, [30] presented a ZigBee-based WSN system for IAQ monitoring in a residential area. They used a CC2530 chip for ZigBee nodes in combination with a 2.4 GHz communication band. It also contains an advanced 8051MCU core and RF transceiver system that can collect IAQ data from field sensors measuring NO_2, SO_2, $PM_{2.5}$, temperature, and humidity.

Adequate monitoring practices can help to reduce the burden of disease with reduced health symptoms. However, WSN-based systems have complicated installation architecture, especially for sensor nodes and field configurations. In this scenario, IoT-based monitoring system architectures are gaining enhanced popularity. These advanced hardware systems allow easy storage of field data on the cloud networks and the data can be accessed on the go, from anywhere, using any digital device. The Wi-Fi-enabled systems make the data transfer process much more convenient. Several hardware environments have been deployed by existing researchers in commercial, residential, industrial, and educational facilities. Refs. [31–35] designed some low-cost and effective IAQ monitoring systems while targeting potential pollutants such as dust, CO_2, NO_2, CO, temperature, and humidity. The researchers also analyzed the impact of ventilation, HVAC systems, and air exchange rate on the building environment; however, the quality of indoor air is observed to be highly influenced by the quality of outdoor air. Marques and Piatra [36] proposed an Internet of Things-based architecture for monitoring indoor air pollution for ambient assisted living applications. The proposed iPAQ system incorporates Bee technology for data transmission and processing along with ESP8266 and Arduino Uno for data acquisition from micro sensors installed in the field environment. The field parameters monitored by researchers included temperature, humidity, CO, CO_2, and light intensity with the help of SHT10, MQ7, T6615, and LDR sensors. The authors also designed a mobile application for real-time access to the IAQ parameters so that relevant information can be accessed by doctors from time to time and relevant support can be extended to the patients at assisted living facilities. The results of this study reveal that such systems can provide viable solutions for improving air quality in the ambient assisted environment which can further reduce the burden of disease by reducing ecological dangers. In a similar manner, Khaliq et al. [37] also designed an IoT-based monitoring system for the nursing home environment by targeting three important IAQ parameters—humidity, temperature, and CO_2. The authors also took outdoor weather conditions into account for the IAQ assessment in the buildings to understand the need for thermal appliances and the airflow requirements on the premises. Results show that higher occupancy of workers and residents in the communal areas was the main reason behind increased CO_2 levels with the window closed. However, opening windows could greatly affect the CO_2 concentrations when the occupancy rate is high.

Mendes et al. in another study [4] analyzed thermal comfort and IAQ in elderly care centers. These researchers explored relevant building characteristics along with environmental variables from 22 different elderly care centers in Portugal to understand the IAQ scenario. They assessed a total of 141 sampling sites for the presence of fungi, total bacteria, CO_2, formaldehyde, CO, VOC, PM_{10}, and $PM_{2.5}$ parameters. At the same time, essential thermal comfort parameters were also analyzed as per ISO 7730:2005 standards. The study shows considerable variation in the pollutant concentrations with seasonal changes. Observations also state that the majority of IAQ parameters are greatly affected by building characteristics including heating ventilation arrangements, insulation type, and window frames. Adequate insulation of walls, ceilings, and windows was found as a reliable solution for improving thermal comfort conditions for elderly care center residents in the winter season.

Unlike stationary sensing systems based on IoT that can access exposure levels in fixed settings, several researchers in recent years have started working on wearable devices. Smart electronic devices these days can be integrated into clothing for pervasive connectivity at the human scale. Smart fitness wearables can sense, track and store environmental data along with biometric data so that the health of elderly patients and disabled people can be monitored actively. With these active monitoring solutions, human comfort can be enhanced by a considerable level. Based on the physiological and active health monitoring applications addressed by several researchers including [38–42], the generic guidelines and recommendations have been developed to address health concerns. Considering the dynamic nature of activities performed by human beings in day-to-day life, several researchers such as [43–45] reviewed the potential of direct assessment of personal exposure level on the go. The portable air quality sensing units with IoT capabilities and online data storage abilities offer a real-time assessment of environmental conditions so that personal exposure estimation can be done in advance. These research studies have also opened doors for establishing a link between environmental pollution and potential health indicators in the human body while taking into account various intrinsic and extrinsic factors. The use of smart devices with IoT sensing capabilities has been more prevalent for elderly patients and disabled people to optimize their health and performance. The data gathered from nursing homes and elderly care centers regarding health outcomes and IAQ conditions can also guide policymakers to develop interventional strategies for building environments.

4. Indoor air pollution management: Methods and approaches

Although monitoring IAQ conditions from the care centers brings all the data relevant to critical scenarios that can affect the health and well-being of individuals. However, intelligent control action is a must to ensure proper IAP management for enhanced health outcomes. Artificial Intelligence offers relevant support in this direction with several theories, technologies, methods, and application systems that can be utilized for expanding and simulating environmental conditions [46]. The main idea behind AI is to enable machines to take automatic actions to prevent human exposure to poor environmental conditions in building environments. AI is being used in almost every sector these days including weather forecasting systems, healthcare management, environmental monitoring, expert system design, and pollutant predictions [47]. In order to identify the risk associated with poor air quality in the building premises, a large number of predictive approaches have been proposed by existing researchers. These methodologies aim to trace the critical interactions between pollutants and their impact on the health of individuals. The predictive monitoring systems can assist in better IAQ management with the ability to take critical actions when required. The existing researchers have utilized a variety of algorithms in this regard from the AI domain, mainly focusing on deep learning and machine learning methodologies. The list of widely used machine learning techniques includes random forest regression, support vector machine, and multilayer perceptron [25,48–52]. However, the deep learning approaches including long short-term memory (LSTM) networks and deep neural networks (DNN) [53–57]. Other than this, there are several mathematical or statistical models that have been tested by existing researchers to assess the relationship between health outcomes and pollutant exposure [58–60]. These authors considered using a generalized additive model, quasi-poison distribution, and poison regression modeling for IAQ assessment. These models helped researchers to assess the hazards associated with potential pollutants such as VOC, O_3, NO_2, SO_2, and $PM_{2.5}$. Several researchers also worked on identifying the impact of IAQ conditions on COVID-19 during pandemic durations. The study conducted by [48] identified the association of aerosol, formaldehyde, and methane with COVID-19 mortality. The researchers [51] determined the effect of ammonia and toluene present in the indoor air on mortality associated with COVID-19. In addition to this, benzene was recognized as another potential chemical substance to increase the risk of COVID-19, especially among patients that are already suffering from chronic diseases such as cancer [61]. This study shows that along with benzene, other heavy elements such as cadmium, arsenic, lead, nickel, benzo-alpha-pyrene, and aromatic hydrocarbon can lead to various kinds of cancers. It is reported that $PM_{2.5}$ exposure can lead to lung cancer. Repeated exposure to NO_2, benzopyrene, and arsenic is further related to intestine disease cases; whereas NO_x caused colorectal and intestine cancer in many individuals.

In earlier years, researchers used several statistical models for IAQ forecasting including the gray model (GM) [62], and autoregressive integrated moving average (ARIMA) [63]. Such models establish links between different environmental variables based on statistical averages and feasibility levels. However, these models could provide limited forecasting accuracy in most cases. The regional characteristics of these statistical models are usually chaotic, complicated, and highly non-linear which reduces the performance accuracy. These drawbacks can be further removed by using AI-based forecasting methods that replicate human vision, reasoning, and learning more efficiently. The generally used techniques for environmental pollution forecasting include Extreme Learning Machine (ELM), Back Propagation Neural Network (BPNN), Wavelet Neural Network (WNN), Long Short-Term Memory (LSTM), Gated Recurrent Unit (GRU), Generalized Regression Neural Network (GRNN), Support Vector Machine (SVM) and Fuzzy Logic Models [64]. Furthermore, the deep learning techniques are capable enough to achieve enhanced forecasting outcomes with extended layers. However, observations state that no single model has shown competent performance for modeling environmental conditions due to individual drawbacks and implementation difficulties. Therefore, the new studies are based on hybrid models that represent an impactful combination of different techniques and algorithms. The list of popular hybrid models includes ensemble empirical mode decomposition combined with least squares SVM [65]; cuckoo search algorithm in combination with principal component analysis and least squares SVM [66]; particle swarm optimization (PSO) in combination with extreme learning machine [67]; PSO combined with BPNN, Adaptive Boosting and Wavelet packet decomposition [68]; BPP combined with random forest and genetic algorithm [69]; SVM with Gray Wolf Optimizer, cuckoo search, and complementary empirical ensemble mode decomposition [70]. The existing researchers investigated that the hybrid models present collective advantages of multiple approaches while eliminating the drawbacks to achieve enhanced prediction outcomes. Several researchers have even utilized AI-based methods for forecasting the health conditions of individuals after critical exposure to degraded air quality. However, the major drawback is that very few studies were conducted specifically in the assisted living environments. Majority of these published papers focus on commercial and residential environments for IAQ assessment. Therefore, the upcoming researchers need to pay more attention towards enhanced environmental assessment for degraded IAQ levels so that health outcomes for the elderly and disabled individuals can be improved. Fig. 1 shows the list of important areas that demand attention of future researchers to design reliable IAQ assessment systems for indoor environments.

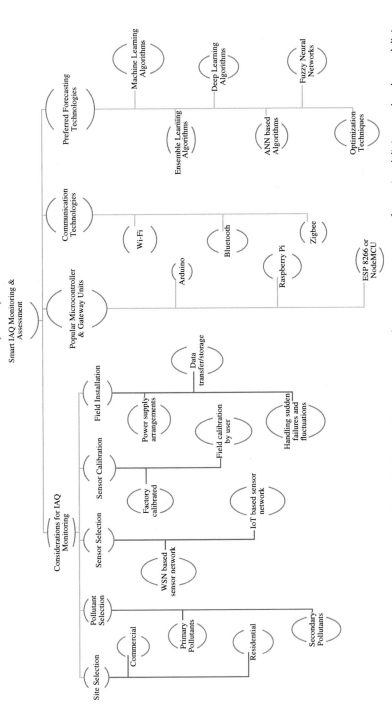

Fig. 1 Focus areas to work upon while establishing IAQ monitoring and assessment systems for assisted living and enhanced living environments.

5. Conclusion

The changing trends in the IAQ monitoring and forecasting domain show that advanced technologies are showing fruitful outcomes in smart building management. A considerable part of WSN-based systems were based on ZigBee-based protocols for communication. The preferred microcontrollers for the IAQ monitoring are Arduino Uno and Raspberry pi. However, the time taken by such WSN-based systems to transfer data from the field environment to the storage platform is considerably high. Also, the battery power management for such systems in case of long-term monitoring becomes critical. Therefore, researchers have now switched to more reliable technologies such as IoT for designing reliable, long-lasting, and cost-effective IAQ monitoring systems. These advanced monitoring systems are capable enough to measure a higher number of pollutants with lesser power consumption, even without demanding any major effort for maintenance. The IoT-based IAQ monitoring systems are named smart solutions for IAQ management in assisted living environments. Furthermore, when IoT-based systems designed are and combined with AI-based approaches, they can lead to reliable prediction outcomes for improved building air quality management. The effective combination of different technologies ensures improved simulation of the environment. The primary requirement for designing these IAQ management solutions is to identify relevant pollutants in assisted living environments to assess hazardous conditions and then choose calibrated sensors to measure these variables. The designed hardware must ensure negligible time delays with lesser power consumption. More importantly, these monitoring systems must be easier to implement in the field environment and the target users must have better ability to interact with these systems.

The major challenge in front of upcoming researchers is to select adequate hardware and software approaches to design smart IAQ solutions. It is also essential to improve the personalization capabilities of smart monitoring systems so that they can be utilized in real-time settings. The upcoming researchers also need to assess the real-world complexity concerns for environmental monitoring. It is necessary to incorporate early warning systems into assisted living facilities for critical IAQ conditions so that impact of dangerous pollutants can be assessed ahead of time. The proposed systems must be utilizable in different locations and building conditions. Therefore, forecasting system training must be done on quality datasets so that the models can avoid underfitting and overfitting conditions. Moreover, such IAQ management systems must be capable enough to reduce the negative impact of IAQ on individuals that spend the majority of their time indoors.

This chapter highlighted the potential methods, approaches, and hardware solutions to design effective IAQ management solutions for assisted living environments. The analysis shows that IoT and AI have relevant potential in this domain. The right combination of these technologies can provide better sensing solutions to design smart IAQ environments. Since there are plenty of pollutants present in the indoor environment, it is important to look after the potential sources in the target premises. In urban areas, the type of pollutants may be different in comparison to the rural sites. Furthermore, IAQ conditions can also change based on lifestyle conditions and routine activities. In order to design efficient systems, it is better to assess the indoor conditions precisely so that a compatible solution can be designed for long-term IAQ management.

References

[1] T.M. Mata, F. Felgueiras, A.A. Martins, H. Monteiro, M.P. Ferraz, G.M. Oliveira, M.F. Gabriel, G.V. Silva, Indoor air quality in elderly centers: pollutants emission and health effects, Environment 9 (2022) 86, https://doi.org/10.3390/environments9070086.

[2] A. Chamseddine, I. Alameddine, M. Hatzopoulou, M. El-Fadel, Seasonal variation of air quality in hospitals with indoor-outdoor correlations, Build. Environ. 148 (2019) 689–700, https://doi.org/10.1016/j.buildenv.2018.11.034.

[3] A. Kankaria, B. Nongkynrih, S.K. Gupta, Indoor air pollution in India: implications on health and its control, Indian J. Community Med. 39 (2014) 203–207, https://doi.org/10.4103/0970-0218.143019.

[4] A. Mendes, S. Bonassi, L. Aguiar, C. Pereira, P. Neves, S. Silva, D. Mendes, L. Guimarães, R. Moroni, J.P. Teixeira, Indoor air quality and thermal comfort in elderly care centers, Urban Climate. 14 (2015) 486–501, https://doi.org/10.1016/j.uclim.2014.07.005.

[5] David, A.P., Russell, M.D., El-Sayed, I.H., Russell, M.S., n.d. Tracheostomy guidelines developed at a large academic medical center during the COVID-19 pandemic. J. Sci. Spec. Head Neck doi:https://doi.org/10.1002/hed.26191.

[6] A. McIntosh, J. Pontius, Chapter 3—Air quality and atmospheric science, in: A. McIntosh, J. Pontius (Eds.), Science and the Global Environment, Elsevier, Boston, 2017, pp. 255–359, https://doi.org/10.1016/B978-0-12-801712-8.00003-2.

[7] S.N. Sinha, P.K. Nag, Air pollution from solid fuels, in: J.O. Nriagu (Ed.), Encyclopedia of Environmental Health, Elsevier, Burlington, 2011, pp. 46–52, https://doi.org/10.1016/B978-0-444-52272-6.00694-2.

[8] A.E. Bauman, J.F. Sallis, D.A. Dzewaltowski, N. Owen, Toward a better understanding of the influences on physical activity: the role of determinants, correlates, causal variables, mediators, moderators, and confounders, Am. J. Prev. Med. 23 (2002) 5–14, https://doi.org/10.1016/S0749-3797(02)00469-5.

[9] M. Almeida-Silva, H.T. Wolterbeek, S.M. Almeida, Elderly exposure to indoor air pollutants, Atmos. Environ. 85 (2014) 54–63, https://doi.org/10.1016/j.atmosenv. 2013.11.061.

[10] A. Baudet, E. Baurès, O. Blanchard, P. Le Cann, J.-P. Gangneux, A. Florentin, Indoor carbon dioxide, fine particulate matter and total volatile organic compounds in private healthcare and elderly care facilities, Toxics 10 (2022) 136, https://doi.org/10.3390/ toxics10030136.

[11] E.L. Pereira, O. Madacussengua, P. Baptista, M. Feliciano, Assessment of indoor air quality in geriatric environments of southwestern Europe, Aerobiologia 37 (2021) 139–153, https://doi.org/10.1007/s10453-020-09681-5.

[12] M. Bentayeb, D. Norback, M. Bednarek, A. Bernard, G. Cai, S. Cerrai, K.K. Eleftheriou, C. Gratziou, G.J. Holst, F. Lavaud, J. Nasilowski, P. Sestini, G. Sarno, T. Sigsgaard, G. Wieslander, J. Zielinski, G. Viegi, I. Annesi-Maesano, Indoor air quality, ventilation and respiratory health in elderly residents living in nursing homes in Europe, Eur. Respir. J. 45 (2015) 1228–1238, https://doi.org/10.1183/09031936. 00082414.

[13] World Health Organization, WHO Global Air Quality Guidelines: Particulate Matter (PM2.5 and PM10), Ozone, Nitrogen Dioxide, Sulfur Dioxide and Carbon Monoxide, World Health Organization, 2021.

[14] C.E. Rodes, P.A. Lawless, G.F. Evans, L.S. Sheldon, R.W. Williams, A.F. Vette, J.P. Creason, D. Walsh, The relationships between personal PM exposures for elderly populations and indoor and outdoor concentrations for three retirement center scenarios, J. Expo. Sci. Environ. Epidemiol. 11 (2001) 103–115, https://doi.org/ 10.1038/sj.jea.7500155.

[15] N. Sousa, B. Segalin, A. Busse, J. Wilson, A. Fornaro, F. Goncalves, Indoor/outdoor particulate matter and health risk in a nursing community home in Sao Paulo, Brazil, Atmospheric Pollution Research 12 (2021), https://doi.org/10.1016/j.apr.2021. 101188.

[16] V.V. Tran, D. Park, Y.-C. Lee, Indoor air pollution, related human diseases, and recent trends in the control and improvement of indoor air quality, Int. J. Environ. Res. Public Health 17 (2020) 2927, https://doi.org/10.3390/ijerph17082927.

[17] D. Krewski, R. Burnett, M. Jerrett, C.A. Pope, D. Rainham, E. Calle, G. Thurston, M. Thun, Mortality and long-term exposure to ambient air pollution: ongoing analyses based on the American Cancer Society Cohort, J. Toxicol. Environ. Health A 68 (2005) 1093–1109, https://doi.org/10.1080/15287390590935941.

[18] Ageing Europe, Looking at the Lives of Older People in the EU, 2020th ed. n.d. [WWW Document], URL (accessed 6.15.23) https://ec.europa.eu/eurostat/web/ products-statistical-books/-/ks-02-20-655.

[19] H.-W. Wu, P. Kumar, S.-J. Cao, Implementation of green infrastructure for improving the building environment of elderly care centres, Journal of Building Engineering 54 (2022) 104682, https://doi.org/10.1016/j.jobe.2022.104682.

[20] M. Viecco, H. Jorquera, A. Sharma, W. Bustamante, H.J.S. Fernando, S. Vera, Green roofs and green walls layouts for improved urban air quality by mitigating particulate matter, Build. Environ. 204 (2021) 108120, https://doi.org/10.1016/j.buildenv. 2021.108120.

[21] S.B. Baker, W. Xiang, I. Atkinson, Internet of things for smart healthcare: technologies, challenges, and opportunities, IEEE Access 5 (2017) 26521–26544, https://doi.org/ 10.1109/ACCESS.2017.2775180.

[22] H. Liu, L. Zhang, K.H.H. Li, O.K. Tan, Microhotplates for metal oxide semiconductor gas sensor applications—towards the CMOS-MEMS monolithic approach, Micromachines (Basel) 9 (2018), https://doi.org/10.3390/mi9110557.

[23] X. Tian, S. Wang, H. Li, M. Li, T. Chen, X. Xiao, Y. Wang, Recent advances in MoS 2-based nanomaterial sensors for room-temperature gas detection: a review, Sensors Diagnostics 2 (2023) 361–381, https://doi.org/10.1039/D2SD00208F.

[24] T.M. Peters, G. Ramachandran, J.Y. Park, P.C. Raynor, Chapter 2—Assessing and managing exposures to nanomaterials in the workplace, in: G. Ramachandran (Ed.), Assessing Nanoparticle Risks to Human Health, second ed., William Andrew Publishing, Oxford, 2016, pp. 21–44, https://doi.org/10.1016/B978-0-323-35323-6. 00002-5.

[25] C.E. Reid, E.M. Considine, G.L. Watson, D. Telesca, G.G. Pfister, M. Jerrett, Associations between respiratory health and ozone and fine particulate matter during a wildfire event, Environ. Int. 129 (2019) 291–298, https://doi.org/10.1016/j.envint. 2019.04.033.

[26] Y. Wu, T. Liu, S. Ling, J. Szymanski, W. Zhang, S. Su, Air quality monitoring for vulnerable groups in residential environments using a multiple hazard gas detector, Sensors 19 (2019), https://doi.org/10.3390/s19020362.

[27] F. Sánchez-Rosario, D. Sánchez-Rodríguez, J.B. Alonso-Hernández, C.M. Travieso-González, I. Alonso-González, C. Ley-Bosch, C. Ramírez-Casañas, M.A. Quintana-Suárez, A low consumption real time environmental monitoring system for smart cities based on ZigBee wireless sensor network, in: 2015 International Wireless Communications and Mobile Computing Conference (IWCMC). Presented at the 2015 International Wireless Communications and Mobile Computing Conference (IWCMC), 2015, pp. 702–707, https://doi.org/10.1109/IWCMC.2015.7289169.

[28] R. Pitarma, G. Marques, B.R. Ferreira, Monitoring indoor air quality for enhanced occupational health, J. Med. Syst. 41 (2016) 23, https://doi.org/10.1007/s10916-016-0667-2.

[29] N. Salman, A. Kemp, A. Khan, C. Noakes, Real time wireless sensor network (WSN) based indoor air quality monitoring system, in: Presented at the IFAC PAPERSONLINE, 2019, pp. 324–327, https://doi.org/10.1016/j.ifacol.2019.12.430.

[30] H. Luo, W. Li, X. Wu, Design of indoor air quality monitoring system based on wireless sensor network, IOP Conf. Ser. Earth Environ. Sci. 208 (2018) 012070, https://doi.org/10.1088/1755-1315/208/1/012070.

[31] C. Balasubramaniyan, D. Manivannan, IoT enabled air quality monitoring system (AQMS) using Raspberry Pi. Indian, J. Sci. Technol. 9 (2016), https://doi.org/10.17485/ijst/2016/v9i39/90414.

[32] L. Barik, IoT based temperature and humidity controlling using Arduino and Raspberry Pi, IJACSA 10 (2019), https://doi.org/10.14569/IJACSA.2019.0100966.

[33] M. Benammar, A. Abdaoui, S. Ahmad, F. Touati, A. Kadri, A modular IoT platform for real-time indoor air quality monitoring, Sensors 18 (2018) 581, https://doi.org/10.3390/s18020581.

[34] M.F.M. Firdhous, B.H. Sudantha, P.M. Karunaratne, IoT enabled proactive indoor air quality monitoring system for sustainable health management, in: 2017 2nd International Conference on Computing and Communications Technologies (ICCCT). Presented at the 2017 2nd International Conference on Computing and Communications Technologies (ICCCT), 2017, pp. 216–221, https://doi.org/10.1109/ICCCT2. 2017.7972281.

[35] Z. Idrees, Z. Zou, L. Zheng, Edge computing based IoT architecture for low cost air pollution monitoring systems: a comprehensive system analysis, design considerations & development, Sensors 18 (2018) 3021, https://doi.org/10.3390/s18093021.

[36] G. Marques, R. Pitarma, An indoor monitoring system for ambient assisted living based on internet of things architecture, Int. J. Environ. Res. Public Health 13 (2016) 1152.

[37] K.A. Khaliq, C. Noakes, A.H. Kemp, C. Thompson, Indoor air quality assessment using IoT-based sensors in nursing homes, in: 2022 14th International Conference On Software, Knowledge, Information Management and Applications (SKIMA). Presented at the 2022 14th International Conference on Software, Knowledge, Information Management and Applications (SKIMA), 2022, pp. 133–139, https://doi.org/10.1109/SKIMA57145.2022.10029568.

[38] C.B. Chan, D.A. Ryan, Assessing the effects of weather conditions on physical activity participation using objective measures, Int. J. Environ. Res. Public Health 6 (2009) 2639–2654, https://doi.org/10.3390/ijerph6102639.

[39] N. Nazarian, J.K. Lee, Personal assessment of urban heat exposure: a systematic review, Environ. Res. Lett. 16 (2021) 033005, https://doi.org/10.1088/1748-9326/abd350.

[40] N. Nazarian, S. Liu, M. Kohler, J.K.W. Lee, C. Miller, W.T.L. Chow, S.B. Alhadad, A. Martilli, M. Quintana, L. Sunden, L.K. Norford, Project Coolbit: can your watch predict heat stress and thermal comfort sensation? Environ. Res. Lett. 16 (2021) 034031, https://doi.org/10.1088/1748-9326/abd130.

[41] L.M. Paulin, N.N. Hansel, Physical activity and air pollution exposures in the urban environment, Am. J. Respir. Crit. Care Med. 194 (2016) 786–787, https://doi.org/10.1164/rccm.201604-0889ED.

[42] M. Romanello, A. McGushin, C.D. Napoli, P. Drummond, N. Hughes, L. Jamart, H. Kennard, P. Lampard, B.S. Rodriguez, N. Arnell, S. Ayeb-Karlsson, K. Belesova, W. Cai, D. Campbell-Lendrum, S. Capstick, J. Chambers, L. Chu, L. Ciampi, C. Dalin, N. Dasandi, S. Dasgupta, M. Davies, P. Dominguez-Salas, R. Dubrow, K.L. Ebi, M. Eckelman, P. Ekins, L.E. Escobar, L. Georgeson, D. Grace, H. Graham, S.H. Gunther, S. Hartinger, K. He, C. Heaviside, J. Hess, S.-C. Hsu, S. Jankin, M.P. Jimenez, I. Kelman, G. Kiesewetter, P.L. Kinney, T. Kjellstrom, D. Kniveton, J.K.W. Lee, B. Lemke, Y. Liu, Z. Liu, M. Lott, R. Lowe, J. Martinez-Urtaza, M. Maslin, L. McAllister, C. McMichael, Z. Mi, J. Milner, K. Minor, N. Mohajeri, M. Moradi-Lakeh, K. Morrissey, S. Munzert, K.A. Murray, T. Neville, M. Nilsson, N. Obradovich, M.O. Sewe, T. Oreszczyn, M. Otto, F. Owfi, O. Pearman, D. Pencheon, M. Rabbaniha, E. Robinson, J. Rocklöv, R.N. Salas, J.C. Semenza, J. Sherman, L. Shi, M. Springmann, M. Tabatabaei, J. Taylor, J. Trinanes, J. Shumake-Guillemot, B. Vu, F. Wagner, P. Wilkinson, M. Winning, M. Yglesias, S. Zhang, P. Gong, H. Montgomery, A. Costello, I. Hamilton, The 2021 report of the lancet countdown on health and climate change: code red for a healthy future, Lancet 398 (2021) 1619–1662, https://doi.org/10.1016/S0140-6736(21)01787-6.

[43] F. Borghi, A. Spinazzè, S. Rovelli, D. Campagnolo, L. Del Buono, A. Cattaneo, D.M. Cavallo, Miniaturized monitors for assessment of exposure to air pollutants: a review, Int. J. Environ. Res. Public Health 14 (2017) 909, https://doi.org/10.3390/ijerph14080909.

[44] L. Spinelle, M. Gerboles, G. Kok, S. Persijn, T. Sauerwald, Review of portable and low-cost sensors for the ambient air monitoring of benzene and other volatile organic compounds, Sensors 17 (2017) 1520, https://doi.org/10.3390/s17071520.

[45] S. Steinle, S. Reis, C.E. Sabel, Quantifying human exposure to air pollution—moving from static monitoring to spatio-temporally resolved personal exposure assessment, Sci. Total Environ. 443 (2013) 184–193, https://doi.org/10.1016/j.scitotenv.2012.10.098.

[46] O.I. Abiodun, A. Jantan, A.E. Omolara, K.V. Dada, N.A. Mohamed, H. Arshad, State-of-the-art in artificial neural network applications: a survey, Heliyon 4 (2018) e00938, https://doi.org/10.1016/j.heliyon.2018.e00938.

[47] D.A. Hashimoto, E. Witkowski, L. Gao, O. Meireles, G. Rosman, Artificial intelligence in anesthesiology: current techniques, clinical applications, and limitations, Anesthesiology 132 (2020) 379–394, https://doi.org/10.1097/ALN.0000000000002960.

[48] N. Amoroso, R. Cilli, T. Maggipinto, A. Monaco, S. Tangaro, R. Bellotti, Satellite data and machine learning reveal a significant correlation between NO2 and COVID-19 mortality, Environ. Res. 204 (2022) 111970, https://doi.org/10.1016/j.envres.2021. 111970.

[49] J. Peng, C. Chen, M. Zhou, X. Xie, Y. Zhou, C.-H. Luo, Peak outpatient and emergency department visit forecasting for patients with chronic respiratory diseases using machine learning methods: retrospective cohort study, JMIR Med. Inform. 8 (2020) e13075, https://doi.org/10.2196/13075.

[50] Z. Ren, J. Zhu, Y. Gao, Q. Yin, M. Hu, L. Dai, C. Deng, L. Yi, K. Deng, Y. Wang, X. Li, J. Wang, Maternal exposure to ambient PM10 during pregnancy increases the risk of congenital heart defects: evidence from machine learning models, Sci. Total Environ. 630 (2018) 1–10, https://doi.org/10.1016/j.scitotenv.2018.02.181.

[51] J.K. Sethi, M. Mittal, Monitoring the impact of air quality on the COVID-19 fatalities in Delhi, India: using machine learning techniques, Disaster Med. Public Health Prep. 16 (2022) 604–611, https://doi.org/10.1017/dmp.2020.372.

[52] J. Shen, D. Valagolam, S. McCalla, Prophet forecasting model: a machine learning approach to predict the concentration of air pollutants (PM2.5, PM10, O3, NO2, SO2, CO) in Seoul, South Korea, PeerJ 8 (2020) e9961, https://doi.org/10.7717/peerj.9961.

[53] J. Ahn, D. Shin, K. Kim, J. Yang, Indoor air quality analysis using deep learning with sensor data, Sensors 17 (2017) 2476, https://doi.org/10.3390/s17112476.

[54] B.S. Freeman, G. Taylor, B. Gharabaghi, J. Thé, Forecasting air quality time series using deep learning, J. Air Waste Manage. Assoc. 68 (2018) 866–886, https://doi.org/10.1080/10962247.2018.1459956.

[55] I.N.K. Wardana, J.W. Gardner, S. Fahmy, Optimising deep learning at the edge for accurate hourly air quality prediction, Sensors 21 (2021) 1064, https://doi.org/10.3390/s21041064.

[56] G. Yang, H. Lee, G. Lee, A hybrid deep learning model to forecast particulate matter concentration levels in Seoul, South Korea, Atmosphere 11 (2020) 348, https://doi.org/10.3390/atmos11040348.

[57] R. Zhang, Y. Tan, Y. Wang, H. Wang, M. Zhang, J. Liu, J. Xiong, Predicting the concentrations of VOCs in a controlled chamber and an occupied classroom via a deep learning approach, Build. Environ. 207 (2022), https://doi.org/10.1016/j.buildenv.2021.108525.

[58] H. Achebak, H. Petetin, M. Quijal-Zamorano, D. Bowdalo, C. Pérez García-Pando, J. Ballester, Trade-offs between short-term mortality attributable to NO2 and O3 changes during the COVID-19 lockdown across major Spanish cities, Environ. Pollut. 286 (2021) 117220, https://doi.org/10.1016/j.envpol.2021.117220.

[59] G. Al Noaimi, K. Yunis, K. El Asmar, F.K. Abu Salem, C. Afif, L.A. Ghandour, A. Hamandi, H.R. Dhaini, Prenatal exposure to criteria air pollutants and associations with congenital anomalies: a Lebanese national study, Environ. Pollut. 281 (2021) 117022, https://doi.org/10.1016/j.envpol.2021.117022.

[60] M. Hadei, P.K. Hopke, A. Shahsavani, A. Raeisi, A.J. Jafari, M. Yarahmadi, M. Farhadi, M. Rahmatinia, S. Bazazpour, A.M. Bandpey, A. Zali, M. Kermani, M.H. Vaziri, M. Aghazadeh, Effect of short-term exposure to air pollution on COVID-19 mortality and morbidity in Iranian cities, J. Environ. Health Sci. Eng. 19 (2021) 1807–1816, https://doi.org/10.1007/s40201-021-00736-4.

[61] N. Tuśnio, J. Fichna, P. Nowakowski, P. Tofiło, Air pollution associates with cancer incidences in Poland, Appl. Sci. 10 (2020) 7489, https://doi.org/10.3390/app10217489.

[62] J. Zhou, J. Zhao, P. Li, Study on gray numerical model of air pollution in Wuan City, in: 2010 International Conference on Challenges in Environmental Science and Computer

Engineering. Presented at the 2010 International Conference on Challenges in Environmental Science and Computer Engineering, 2010, pp. 321–323, https://doi.org/10.1109/CESCE.2010.30.

[63] H. Liu, G. Yan, Z. Duan, C. Chen, Intelligent modeling strategies for forecasting air quality time series: a review, Appl. Soft Comput. 102 (2021) 106957, https://doi.org/10.1016/j.asoc.2020.106957.

[64] S. Subramaniam, N. Raju, A. Ganesan, N. Rajavel, M. Chenniappan, C. Prakash, A. Pramanik, A.K. Basak, S. Dixit, Artificial intelligence technologies for forecasting air pollution and human health: a narrative review, Sustainability 14 (2022) 9951, https://doi.org/10.3390/su14169951.

[65] Y. Bai, B. Zeng, C. Li, J. Zhang, An ensemble long short-term memory neural network for hourly PM2.5 concentration forecasting, Chemosphere 222 (2019) 286–294, https://doi.org/10.1016/j.chemosphere.2019.01.121.

[66] W. Sun, J. Sun, Daily PM2.5 concentration prediction based on principal component analysis and LSSVM optimized by cuckoo search algorithm, J. Environ. Manag. 188 (2016) 144–152, https://doi.org/10.1016/j.jenvman.2016.12.011.

[67] W. Sun, C. Huang, Predictions of carbon emission intensity based on factor analysis and an improved extreme learning machine from the perspective of carbon emission efficiency, J. Clean. Prod. 338 (2022) 130414, https://doi.org/10.1016/j.jclepro.2022.130414.

[68] H. Liu, K. Jin, Z. Duan, Air PM2.5 concentration multi-step forecasting using a new hybrid modeling method: comparing cases for four cities in China, Atmos. Pollut. Res. 10 (2019) 1588–1600, https://doi.org/10.1016/j.apr.2019.05.007.

[69] S.-Q. Dotse, M.I. Petra, L. Dagar, L.C. De Silva, Application of computational intelligence techniques to forecast daily PM10 exceedances in Brunei Darussalam, Atmos. Pollut. Res. 9 (2018) 358–368, https://doi.org/10.1016/j.apr.2017.11.004.

[70] S. Zhu, X. Qiu, Y. Yin, M. Fang, X. Liu, X. Zhao, Y. Shi, Two-step-hybrid model based on data preprocessing and intelligent optimization algorithms (CS and GWO) for NO2 and SO2 forecasting, Atmos. Pollut. Res. 10 (2019) 1326–1335, https://doi.org/10.1016/j.apr.2019.03.004.

About the authors

Jagriti Saini is the Founder and Owner of "Eternal RESTEM"—a Startup working in the direction of overall student growth and welfare at multiple levels. Jagriti Saini holds a Diploma in Electronics and Communication Engineering (2010) from GPW Kandaghat and completed her B. Tech in Electronics and Communication Engineering (2013) from HPU. She received a Master's Degree in Electronics and Communication Engineering from the National Institute of Technical Teacher's Training and Research (NITTTR),

Chandigarh (Panjab University), India (2017). She was awarded a Gold Medal for securing the highest percentile in the entire university during her Master's Degree. Jagriti completed her PhD from NITTTR, Chandigarh (2022) with affiliation to Panjab University. She received INSPIRE fellowship from the Department of Science and Technology (DST), India for carrying out her research work. She also completed her Post Graduate Certificate Programme on Data Science and Business Analytics from Purdue University (USA) in collaboration with IBM and Simplilearn (2023). Her current research interests include Data Science, Artificial Intelligence, the Internet of Things, Environmental Monitoring, Indoor Air Quality Monitoring and Prediction, Healthcare Systems, e-Health, and Autonomous Systems. She published more than 30 papers in reputed peer-reviewed international journals and conferences. Other than this, she is a frequent reviewer of journals and international conferences and is also working on several edited book projects.

Maitreyee Dutta was born in Guwahati, India. She received a BE degree in electronics and communication engineering in 1993 from Guwahati University and was Gold Medalist in the same year. She obtained an ME degree in electronics and communication engineering, and a PhD degree in the faculty of engineering from Panjab University. She is currently a Professor and Head of Information Management and Emerging Engineering and a Joint Professor in the Computer Science and Engineering Department, at the National Institute of Technical Teachers' Training and Research, Chandigarh, India. She has more than 22 years of teaching experience. Her research interests include the Internet of Things, security of data, IP networks, Internet, authorization, data privacy, Public Key encryption, pattern clustering, cloud computing, and data compression. She has more than 100 research publications in reputed journals and conferences. She completed two sponsored research projects—Establishment of Cyber Security Lab—funded by the Ministry of IT, Government of India, New Delhi, amounting to Rs. 45.65 lakhs, and Establishment of

Advanced Cyber Security Lab sponsored by MeitY, New Delhi amounting to 62 lakhs. One sponsored project Securing Billion of Things-SEBOT funded by All India Council of Technical Education, New Delhi of amount Rs. 14.98 lakhs is in progress.

Gonçalo Marques holds a PhD in Computer Science Engineering and is a senior member of the Portuguese Engineering Association (Ordem dos Engenheiros). He is currently working as Assistant Professor lecturing courses on programming, multimedia, and database systems. Furthermore, he worked as a Software Engineer in the Innovation and Development unit of Groupe PSA automotive industry from 2016 to 2017 and in the IBM group from 2018 to 2019. His research interests include the Internet of Things, Enhanced Living Environments, machine learning, e-health, telemedicine, medical and healthcare systems, indoor air quality monitoring and assessment, and wireless sensor networks. He has more than 80 publications in international journals and conferences, is a frequent reviewer of journals and international conferences, and is also involved in several edited book projects.

CHAPTER FIVE

Usability evaluation for the IoT use in Enhanced Living Environments

Hana Kopackova ⓘ and Miloslav Hub ⓘ

University of Pardubice, Faculty of Economics and Administration, Institute of System Engineering and Informatics, Pardubice, Czech Republic

Contents

Abstract

The evaluation of usability represents an essential step in designing and developing new technologies and comparison of existing ones. It allows researchers to identify any user experience issues with the product, decide how to resolve them, and ultimately determine if the product is sufficiently usable by specific users in a specific context of use. Enhanced Living Environments (ELE) aim to prolong the independent life of supported persons and to reduce dependence on intensive personal care to a minimum. ELE incorporates the latest developments related to the Internet of Things to offer better help and support for all stakeholders; customers (elderly or persons with disabilities), physicians, caregivers, family members and friends, and service providers. This chapter aims to bring a comprehensive and systematic view of the usability evaluation methods exploited during the design of ELE technologies. The chapter brings two views on this topic: (1) identification of the characteristics of ELE research articles incorporating usability evaluation and (2) systematization of the methods, procedures, and instruments being used.

Advances in Computers, Volume 133
ISSN 0065-2458
https://doi.org/10.1016/bs.adcom.2023.10.004

127

1. Introduction

Developments in science and technology (both medical and non-medical) are changing the landscape of home care solutions that make staying at home feasible for the larger share of the population. Self-sufficiency and independent living are the wishes and needs not only of young and healthy people but also of the elderly or people with certain limitations. A national survey of housing and home modification issues [1] proved that people (in the US) prefer to be able to remain in their homes as they grow older. A similar perception is articulated by the study of World Health Organization Europe, given the European environment [2]. People's wishes are only one side of the coin.

Developments in home care are also inevitable as the population is aging and healthcare and institutional care expenditures are rising. United Nations study [3] from 2019 forecasts that "by 2050, 1 in 6 people in the world will be over the age of 65, up from 1 in 11 in 2019." The projected number of elderlies is 1.5 billion persons in 2050. Although aging is fastest in Eastern and South-Eastern Asia, Latin America, and the Caribbean, almost every country in the world is experiencing growth in the size and proportion of older persons in their population.

Aside from the number of older people, there is the fundamental care they need according to their health problems. On the other hand, we can see a decline in the availability of family caregivers. Providing older persons with services that support their ongoing health and social needs in hospitals and other institutions is expensive and burdens the public budget [4]. Moreover, older adults are not the only source of higher costs on health care; even people of productive age burden the health system due to the development of chronic diseases (cardiovascular diseases, cancers, respiratory diseases, and diabetes).

Four facts support the urgency of the ELE topic:

1. The number of older persons and chronically ill is growing every year.
2. Caring for these groups is challenging concerning the costs and necessary help of caregivers.
3. Informal caregivers (family, friends) are often unavailable, meaning these people must stay in institutional facilities.
4. The majority of people wish to stay at home instead of institutional facilities.

This chapter aims to extend ELE research by identifying and systematically reviewing studies focused on usability evaluation in the ELE context. Studies included in the review must meet several conditions: studies must be fully described, use IoT technology, and none of them had been previously evaluated by Bastardo et al. [5]. This systematic literature review should reveal characteristics of ELE research incorporating usability evaluation and systematize these usability evaluation methods, procedures, and instruments.

The rest of the chapter is organized as follows. The next two sections introduce theoretical concepts employed in our research. Namely, it is the Internet of Things (IoT) and usability evaluation. The methodological framework then clearly explains the whole process of literature review. The results are presented in the next section. Finally, the last section gives the conclusion and the proposal for future research activities.

1.1 Internet of Things in the environment of enhanced and assisted living

The concept of home environments equipped with advanced technologies that make life easier for its inhabitants has become popular under the umbrella of different terms and definitions. Enhanced Living Environments (ELE) and Ambient Assisted Living (AAL) are primarily used concerning home care for elderly and persons with disabilities. In contrast, the term Smart homes is mostly perceived as a fancy label used to describe buildings equipped with smart gadgets. Although the target group differs, both views share the same operating principle. The technology behind the scenes embraces developments in the Internet of Things (IoT) field.

IoT is a new trend in the control and communication of objects of everyday use with each other or humans, mainly through wireless data transmission technologies and the Internet. The Internet of Things aims to connect devices, systems, and services to provide more data that can be translated into information and relevant knowledge. This way, IoT systems can make decisions and carry out activities autonomously.

The conceptual framework of IoT mainly uses layered models at different levels of detail. The basic architecture model, which has three layers, comes from the wireless sensor network concept [6,7] and comprises the Perception layer, the Network layer, and the Application layer. The second model adds the Processing layer as the processing logic (data integration, analysis, and storage) can be independent of the particular application and shared through different applications [8,9]. The third model, which comprises five layers, divides the Application layer into two separate parts. The Application layer in this model is responsible for communication among the gateway, Internet, and application, whereas the Business layer defines the goal of the service and its management and form of the presentation [10,11]. The last model, which is the most detailed, comprises seven layers [12]. The logic of the Middleware layer is divided into three components: Edge computing (data element analysis), Data Accumulation (storage), and Data Abstraction (integration, filtering). A graphical comparison of different models is depicted in Fig. 1.

The Perception layer, which is the same for all models, includes sensors and actuators uniquely identifiable in the network. The purpose of sensors is to collect data (sensing) through different means (bar code labels, RFID tags, GPS, camera, sensors for noise, CO_2, and temperature, etc.) and send them (communication) through the Network layer for further processing. On the other hand, actuators receive information requiring changes in the setting of controlled devices. In addition, different actuators can be used: hydraulic, pneumatic, electrical, mechanical, thermal, or magnetic.

Definition of the Network layer is similar for all models, addressing the transfer of gathered data from the sensors to the point of processing and back to actuators. Network architectures depend on battery consumption and data range [13]. The shortest distance and low power consumption represent Wireless Personal Area Network (WPAN), which includes proximity data transfer as RFID or NFC and networks like Bluetooth. A Wireless Local Area Network (WLAN) includes ZigBee, Z-Wave, 6LoWPAN, or

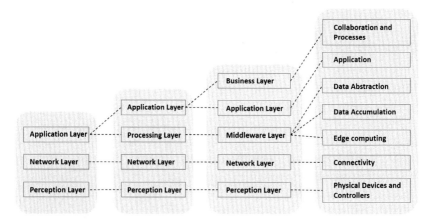

Fig. 1 Conceptual models of IoT—a variety of detail.

Wi-Fi on a slightly larger wireless network area scale. A new type of network, the LPWAN (Low Power Wide Area Network), has been developed specifically for the Internet of Things. This network technology group is designed to wirelessly communicate small data packets on low transmission data rates over relatively long distances using lower power than standard network technologies (Sigfox, LoRaWAN, NB-IoT). On a larger scale, mobile communication technologies (2G, 3G, 4G, 5G, LTE) can transmit data [14,15].

Data processing represents the most critical part of the IoT system that transforms the data into valuable knowledge [10,16]. Processing layer encompasses device management support, security, data preparation phase (accumulation, verification, abstraction, integration, storage), and data analysis (machine learning, predictive analytics, data mining, artificial intelligence). While the purpose of data processing remains the same over time, we can observe a shift in the place where this processing takes place.

Cloud computing can provide data processing, moving all data from the point of gathering onto the Internet; nevertheless, this approach can suffer some disadvantages (data security threats, performance issues, or operational costs). Therefore, another approach incorporates Edge or Fog computing into the processing phase. Fog and Edge computing are network and system architectures that attempt to collect, analyze, and process data from the Perception layer more efficiently than traditional cloud architecture [17,18]. The difference between them is in the proximity to sensors. Edge computing means incorporating logic and processing power right into sensing devices [19,20] to make the response as fast as possible using PLCs (programmable logic controllers), PACs (programmable automation controllers), or EPICs (edge programmable industrial controllers). On the other hand, fog computing is closer to cloud computing, being the layer in between the edge and the cloud. The purpose of Fog computing is to be the nod that helps in data analytics and filtering important information from data collected by sensors and sending filtered data to the cloud.

The Application layer is responsible for providing services to the end user (Web Services, User Interface (UI) of specific applications using the gathered data). The Application layer uses different protocols that enable gateway, Internet, and application communication [21,22]. Application layer protocols are founded on TCP and UDP. The most popular Application layer protocols are Message queue telemetry transport (MQTT), Extensible messaging and presence protocol (XMPP), Advanced message queuing protocol (AMQP), or Constrained application protocol (CoAP).

The Business layer [23] offers system administrators tools to manage the overall functionality of the IoT platform—cooperation among all devices, nods, and applications, user security and business process management. The use of IoT is widespread in the majority of industries and even the whole society. Examples can be found in Industry 4.0, smart cities, smart healthcare, self-driving cars, smart homes, and so on. The list of business models is already exhaustive, but it is expanding every year.

The focus of this chapter is the use of IoT in ELE or AAL, which represents the business model combining the technology for smart homes and smart healthcare to deliver valuable solutions for elderly and persons with disabilities. Such solutions can be divided, according to Ref. [5], into four groups: (1) Secure and supportive environment, (2) Healthcare provision, (3) Healthy lifestyle, and (4) Social involvement and active participation.

The primary goal of these systems is to help people who would need institutional care as they cannot take care of themselves. IoT can change this situation by using sensors gathering data about residents that are automatically assessed by software or by a caregiver connected to this system. Actuators can display some information, warn residents, or act (e.g., switch off the gas supply if a leak is detected, or prepare medications at the given time and warn people). IoT solutions can also be divided according to the action expected from the user. Active IoT solutions require user activity to gather the data. For example, a blood pressure monitor [24] is an active IoT solution as it requires the user to start measuring and save the data. Moreover, this solution allows the user to monitor developments using statistics for longer period of time. On the other hand, fall prevention systems [25] are passive, which means it does not require any activity from the user side. Instead, the system monitors the user's behavior, and if it evaluates it as risky, it triggers an alarm and calls for help.

1.2 Usability evaluation

Usability evaluation dates back to the 90s when the book Usability engineering by Jacob Nielsen [26] was first published. In this book, the author described existing engineering practices in the field that had been considered more of a field of psychology. It was a time when the number of competing software products grew with the growing number of end users rapidly, and users were not forced to use a software product that did not meet their requirements simply because there was no alternative. At that time, developers had to admit that end users must be the focus of software development. Therefore, new fields that deal with the human-computer relationship began to emerge. They were known as computer–human interaction (CHI), human–computer interaction (HCI), user-centered design (UCD), human factors (HF), and ergonomics. As the usability of a user interface became important, usability engineering gradually emerged from these fields and became an integral part of software engineering [26].

The goal of usability engineering is not to defend the importance of usability, it has been proven many times in the past, but to create user interfaces that will be usable and used by end users [27] through measurement, evaluation, and usability improvement [26]. The subject of interest in usability engineering is often the usability of software applications. However, it can also be, for example, user interfaces of remote controls, car dashboards, or scanner control.

All current software quality models, whether McCall [28], FURPS [29] or FURPS+ [30], or ISO/IEC 9126 [31], consider software usability as one of the key features. According to ISO 9241 [32], usability is understood as "*Efficiency, performance, and satisfaction with which a specific user achieves specific goals in a specific environment.*" Efficiency is the precision and completeness with which a specific user can perform specific tasks in a particular environment. Performance represents the effort to achieve these goals with the required accuracy and completeness. Satisfaction represents the convenience and acceptability of the product for end users and other users affected by its use. A different view of usability offers the MUSiC (Metrics for Usability Standards in Computing), which is the metric based on acceptability and learnability. MUSiC [33] defines usability as: "*Ease of use and acceptability of a system or product by a group of users performing tasks in a given environment, where ease of use affects performance and satisfaction, acceptability affects whether or not the product is used.*" Another current definition of usability can be found in [34], which defines usability as: "*Usability is the extent to which users can use computer systems to achieve specified goals efficiently and effectively while supporting a sense of satisfaction in a given context.*"

Usability cannot be understood as a one-dimensional feature of the user interface but as composed of several different components. The literature usually lists the following five most important components of usability [26,35]:

- learnability—fast and sufficiently accurate creation of a mental model of the user interface,
- efficiency—providing expected and desired performance,
- memorability—the ability to maintain a mental model of the user interface,

- satisfaction—subjective individual feeling associated with interaction with the user interface,
- errors—minimization of incorrectly performed activities in the user interface.

Usability engineering has a large number of methods nowadays. By Refs. [26,36,37], usability methods can be divided according to the data source they use, namely, user-oriented usability testing methods, expert review-based usability evaluation methods, and usability evaluation methods based on models. User-oriented usability methods use end-user representatives (participants) who perform assigned tasks on a prototype of the software user interface. Subsequent analysis of the data obtained through these experiments (average number of errors, the time required to complete the task) identifies usability weaknesses [38,39]. Nielsen [26] states that real user testing is a primary and irreplaceable method, as it provides immediate information on how users use the products and what specific problems they encounter with the user interfaces being tested. Expert-based usability evaluation methods are based on reviewing user interface usability aspects by an expert (evaluator) concerning a set of principles, either explicitly stated or in the form of expert knowledge [34]. Finally, usability evaluation methods based on models such as GOMS [40] analysis or UIDE [38] analysis allow predicting usability through the user interface and/or user model quickly, inexpensively, and without experts or participants.

Usability engineering methods can be applied at different stages of product development when there are different needs, these methods can be divided according to their purpose into formative and summative methods (see Fig. 2).

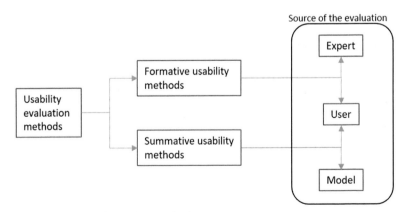

Fig. 2 Classification of usability methods.

Formative usability methods are used to find usability problems with an existing product user interface or its prototype that the development team must fix to make it more usable. These methods' outputs are quantitative and are usually used in the early stages of product development. Examples of these methods are thinking-aloud protocol [26], heuristic evaluation [41], and cognitive walkthrough [42]. In the field of formative usability engineering methods, users and experts tend to be the data source; no known models could be used for these purposes. The use of user-sourced methods dominates over expert-based methods.

Summative usability methods describe how well a product user interface performs, often compared to some benchmark, for example, a prior version of a product or a competitor's product. The output of these methods is a quantitative value(s) that can be easily compared. These methods are usually used in the late stages of product development or in the case of the final product. Examples of these methods are System Usability Scale (SUS) [43], Usability Magnitude Estimation [44], or Expected Usability Magnitude Estimation (Expected UMA) [45]. Summative usability engineering methods obtain their data through users and models; existing literature does not report summative methods using experts as a data source. The number of methods using users dominates methods using models, which are currently not applied very often.

2. Methodological framework

This section introduces a research design based on the systematic literature review we employed to fulfill the aim of the article. The aim was to bring a comprehensive and systematic view of the usability evaluation methods exploited during the design of ELE. We had to specify the scope of the review in broader terms as the ELE is not the only possible title for the concept of our interest. We tried not to omit thematically correct articles using different terminology. The scope of the review covers all possible business models combining the IoT technology for smart homes and smart healthcare to deliver valuable solutions for elderly and persons with disabilities allowing them to stay at home instead of institutional care. To our knowledge, only Bastardo et al. [5] conducted a systematic literature review on usability evaluation methods in an Ambient Assisted Living Environment. Their article revealed 44 studies covering this topic. Nevertheless, they also pointed out that the concept of AAL has a poor expressiveness outside Europe, which was proved by only three studies originating outside Europe. This fact motivated us to enhance the query expressions in our research to cover more papers (preferable outside Europe) within the scope.

The electronic literature search was performed in May 2022 through Web of Science and included all the references published before this day. We defined four query expressions: two of them narrow ("enhanced living," "ambient living"), covering precisely the topic with given terminology, and two of them broad ("assisted living," "ubiquitous computing") offering a pool of papers focusing on assistive technology using any device, in any location, and any format.

Fig. 3 depicts the whole process of the review. In the first step, we defined keywords and gathered and preprocessed data. The total number of sources was 13,387. Adjusting for duplication reduced this to 13,237, and excluding papers without abstracts further reduced the number to 12,106. In the next step, we searched abstracts for two queries: "Usability" or "User experience." Four hundred sixty-five papers met these criteria. The last step in the preprocessing stage, which reduced the number to 460, was excluding non-English papers.

The second step included a careful reading of abstracts and a selection of papers that were thematically correct. Only papers dealing with technological solutions intended for the independent stay of elderly or disabled persons at home were included. This step reduced the number of papers to 108. In the next step, we compared papers evaluated in Ref. [5] with our

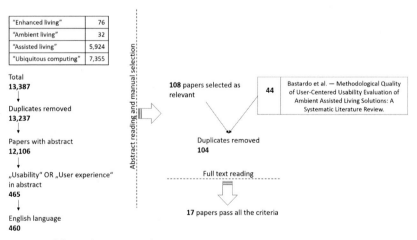

Fig. 3 Workflow of systematic literature review.

selection and eliminated four duplicates. A shortlist of 104 papers formed the basis for the whole-text review.

In the final step, we read all papers to find those that correspond to our three screening criteria:

1. The paper contains a usability study, which is fully described. This criterion eliminated papers that described some ELE solutions, but the usability was mentioned only in the context of necessary future research or as a reference to other papers.

2. The ELE solution tested in the paper uses IoT technology. Many types of IoT technology made the application of this criterion complicated. Therefore, we narrowed the specification as the use of technology that gathers and processes data using objects of everyday use. This criterion eliminated papers testing devices offering services without gathering data about their use, for example, the pills reminding system, which works as an alarm clock. On the other hand, if this system gathers the data and provides some other service, such as contacting the caregiver or displaying statistics, we selected this paper for processing.

3. The study describes a unique project not evaluated in Ref. [5]. During the review process, we discovered that, in some cases, more papers describe the same project from different perspectives. As this chapter aims to find specific characteristics of ELE research, we decided to eliminate these overlapping papers.

At the end of the review process, we got a final selection of 17 papers that included unique usability studies of IoT technology in the ELE environment. These papers formed the basis for the evaluation from two viewpoints: (1) characteristics of ELE research and (2) types of usability methods. We also compared our results to those described in Ref. [5].

3. Results

Selected publications cover the period from 2003 to 2021. The highest publication activity was mentioned in 2014 and 2021 with three publications, which shows, in accordance with Ref. [5], that the interest in this topic has been increasing in recent years. We found that most of the research was done in Europe according to the origin of selected publications. Although we have broadened keyword search to find papers that use different terminology but deal with the given topic, we found that usability testing of ELE environment is mainly European topic (see Fig. 4), even though people are getting older all over the world. The reason can be found

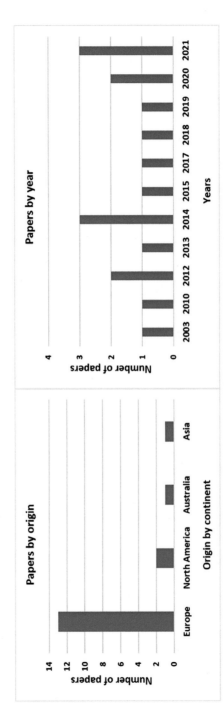

Fig. 4 Comparison of origin and year of publication of selected papers.

in the number of projects financially supported by the European Union and member states. For example, Active Assisted Living (AAL) Association, founded in 2007 by 14 member organizations, supports and finances the research and development of technologies and services for older adults. This program is currently co-funded by national and regional public funding agencies of 18 European countries and the European Commission (through Horizon 2020). To date, this agency has supported 309 projects worth 440 million euros.

3.1 Characteristics of ELE research articles with usability evaluation

The content and the thematic focus of selected papers differ as the ELE topic is vast and covers different domains as assistive technology, human–computer interaction, health and well-being or ethics and privacy issues. In this chapter, we follow the division published by Bastardo et al. [5] to make the research comparable. Most papers according to this division span into Healthcare provision [24,46–51]. A secure and supportive environment was the topic of five papers [52–56]. Social involvement and active participation domain comprise of papers [57,58], and part of the paper [51]. Finally, three papers dealt with a Healthy lifestyle, covering the topic of physical and cognitive games [59–61].

A closer look at the articles in the healthcare domain (see Table 1) shows that the main topic is the collection of data on the user's vital functions and sharing of these data with caregivers. Data gathering devices were primarily used for monitoring blood pressure, glucose, and oxygen saturation [24,46–48,51]. Medication management is also an important topic in this field as most seniors need to take some pills and attend to physicians regularly. Reminding systems equipped with feedback ensure that the medication is taken at the right time. Data about medication use can be accessed and monitored remotely. In case of a problem, caregivers can be warned [49,50].

Smart technology in homes of elderly provides a secure and supportive environment through different means (see Table 2). The first type of research in this domain focuses on the quality of interface of the system used by caregivers to monitor the daily life activities of the elderly. Testing, in this case, approached a different target group (caregivers) than in previous studies [52]. Forgetfulness is a common symptom of aging, but it can also be very annoying. Soar and Symonds [53] solved this problem with RFID tags placed at losable objects. Another example is controlling the temperature in the

Table 1 Healthcare provision domain—Function (Blood Pressure, Glucose Level, Oxygen Saturation, Heart Rate, Medication Management), Data access (user or caregiver).

References	Project name	Technology	B	G	O	H	M	Data access
[24]	MoMo	Biometric and mobile device, Bluetooth, mobile app	x	–	–	–	–	U/C
[46]	IDEAT	Biometric device, web-based Diabetes Manager, videoconferencing, email	x	x	–	–	–	U/C
[47]	eCAALYX	Biometric device, Set-Top-Box, Caretaker Server, Auto-configuration Server for the Home System	x	–	x	x	–	U/C
[48]	Self-management System	Paper prototype of computer-based app	x	x	x	–	x	U/C
[49]	ACHO	Mobile application, smartphone, voice assistant, Bluetooth	–	–	–	–	x	C
[50]	EMMA	Blister cards in the drawer, cellular modem, LAN connection, touch screen	–	–	–	–	x	U/C
[51]	Ella4Life—Emma assistant	Mobile application, smartphone, Bluetooth, cloud, sensors	x	x	x	x	x	U/C

apartment [54] with the smart thermometer. Sensors that control the movement of the monitored person enabling remote monitoring of fall or inactivity was discussed in Ref. [55]. IoT technology can also control lights and automate selected activities such as opening and closing doors, curtains, and shutters in the apartment's room [56].

Technologies for Social involvement help people to overcome isolation and loneliness. Such technologies can be embedded in the computer-type

Table 2 Secure and supportive environment domain—Purpose (Loss prevention, Thermostat, Lightning, Doors/Windows/Curtains, Fall prevention, Daily activity), Data access (user or caregiver).

References	Project name	Technology	LP	T	LI	DWC	FP	DA	Data access
[52]	Loved-1 UI	Paper prototype	–	–	–	–	–	x	C
[53]	Aura Object Location	RFID tag, mobile app, PDA	x	–	–	–	–	–	U
[54]	Manuela	Smart thermostat, virtual assistant	–	x	–	–	–	–	U
[55]	TeleStiki	Mobile app	–	–	–	–	x	x	C
[56]	DOMHO	IoT sensors mobile app, intelligent video cameras	–	x	x	x	x	–	U/C

Table 3 Social involvement and active participation domain—Purpose (Video conferencing, e-mail, radio, physical delivery, news, calendar, medication, album, games), Data access (user or caregiver).

References	Project name	Technology	VC	E	R	PD	CA	M	A	G	Data access
[51]	Ella4Life—Anna assistant	Tablet (C++/C99), websockets	x	x			x	x	x	x	U
[57]	AALFred assistant	Multimodal interaction	x				x				U
[58]	Kompai robot	Assistive mobile robot platform	x		x	x	x	x	x	x	U/C

device or more "human" type as a social robot. Technical solutions in the social involvement and active participation domain can be divided into two groups (see Table 3). The first category belongs to software solutions that do not need specialized hardware equipment. These solutions are cheaper and thus more common. Users can choose the type of device they will use (PC, tablet, smartphone, smart TV) and mostly decide what types of interaction modalities are the most suitable for them (visual, voice interaction, or gestures). Our review found two representatives of this category [51,57].

The second category includes social robots that imitate humans by their look and ability to move. Assistive robots may be assigned functions such as fetching things, reminding, helping with daily activities, etc. Only one paper discussed this type of solution [58].

The last research domain covers maintaining a healthy lifestyle (see Table 4). Physical and cognitive activity is vital at all ages; however, at a senior age, there is less opportunity to train both physical and cognitive functions. Therefore, solutions that allow seniors to train these functions at home without needing to attend the specialized facility are welcomed and should be the focus of ELE research. Nevertheless, our study shows that only three papers solved this problem [59–61].

To summarize the first part of our research, the highest attention in ELE research, which incorporates usability studies, is devoted to healthcare provisions. Most studies in this domain concentrated on gathering vital data and medication management. On the other hand, the least attention is focused on a healthy lifestyle and social involvement.

3.2 Systematization of the methods, procedures, and instruments being used in ELE research

When testing and evaluating the usability of ELE, the entire spectrum of usability engineering methods was not used, but only a few that are often repeated in individual studies (see Table 5). These are often user testing, questionnaire surveys, interviews, and the SUS scoring method. A common way is to test ELE technology by the end user and then fill out a questionnaire in which the end user expresses his feelings about the use of the given technology, for example Refs. [24,47,54]. Of course, one can argue that not

Table 4 Healthy lifestyle domain—Purpose (Physical training, Cognitive training, Game, Daily activity task), Data access (user or caregiver).

References	Project name	Technology	PT	CT	G	DAT	Data access
[59]	GameUp extragames	MS Kinect, TV	x	x	x	–	U
[60]	Bal-App	Portable cycle ergometer, Apple IPad 2, Bal-App	x	x	x	–	U
[61]	CogWatch	Speakers, vibro-tactile actuators, visual displays, watch, PC monitor, MS Kinect, Bluetooth	x	x	–	x	U/C

Table 5 Usability engineering methods applied in practice—Test method (Performance, Observation, Think aloud, Wizard of Oz); Inquiry method (Interviews, Scales, Questionnaires); Test environment (Institutional site, Living lab, Participant home, Research lab).

References	Test				Inquiry			Test environment			
	P	O	T	W	In	S	Q	I	L	P	R
[24]	–	–	–	–	–	–	x	–	–	x	–
[46]	x	x	–	–	–	–	–	–	–	x	–
[47]	–	–	–	–	–	–	x	–	–	x	–
[48]	x	x	x	–	–	–	x	x	–	–	–
[49]	–	–	–	–	x	x		–	–	x	–
[50]	x	x	–	–	–	x		–	x	–	–
[51]	–	–	–	–	–	–	x	–	–	x	–
[52]	–	–	–	–	–	–	x	–	–	–	x
[53]	x	x	–	–	–	–	–	–	–	–	x
[54]	–	x	–	–	–	–	x	–	–	–	x
[55]	x	x	x	–	x	x	x	x	–	–	x
[56]	–	–	–	–	x	x	x	–	x	–	–
[57]	x	x	x	–	x	–	x	–	x	x	–
[58]	x	x	–	x	x	–	x	x	x	x	–
[59]	–	x	–	–	x	–	–	x	–	–	–
[60]	–	–	–	–	x	x		x	–	–	–
[61]	x	x	–	–	–	x	x	–	–	–	x

all usability engineering methods can be used in ELE usability testing, which has its specifics. However, many methods could be used either directly or after minor modification.

An example is the heuristic evaluation, which is relatively cheap and fast, does not require the participation of end users, and yet has only been used once in the evaluated studies [48]. It should also be emphasized that in none of the evaluated studies, the choice of the given method is explicitly justified. None of the monitored studies dealt with comparing the performance and appropriateness of different usability engineering methods by simultaneously applying different methods to the same problem and comparing the outputs of these methods to each other, as is quite common in the field of usability engineering, for example Refs. [62–64]. Instead, these studies choose the optimal method (and its settings) in a given situation.

When testing and evaluating the usability of ELE technology, different types and numbers of participants were used (see Table 6). A specific feature of the AAL field is a relatively broad spectrum of stakeholders with different characteristics (abilities, knowledge, skills), which must be considered when applying usability engineering methods (for example, seniors

Table 6 Characteristics of the participants.

References	Participants Type	No	Participants characteristics description
[24]	Users	23	Adults—10 retired, 13 (age 25–72)
[46]	Potential and actual users	25	Persons with diabetes
[47]	Users and caregivers	9+	Older adults—9 (age 60+) and not a specified number of caregivers
[48]	Experts, users	50	Adults—(age 55+) with a diagnosis of any chronic disease, normal vision or corrected-to-normal vision, no cognitive or physical impairment, and the ability to read Traditional Chinese
[49]	Users	10	Older adults—(age 65+), 5 male, 5 female with no cognitive or sensory impairment, no critical medication
[50]	Users	19	White female (age 80+), higher education
[52]	Informal caregivers (family)	130	Adults 18+ years having experience with caring for the elderly (66 male, 64 female)
[51]	Users (seniors)	36	Older adults—23 (age 55–65); 10 (age 66–75); 3 (age 75+)
[53]	No relationship	5	Healthy university staff—(age 33, 34, 60, 61, 49)—1 male, 4 female
[54]	Users	40	20 young students (age 20–40), 20 older adults (age 50–80)
[55]	Informal caregivers	21	Adults—(age 19–38), 10 male, 11 female

Table 6 Characteristics of the participants.—cont'd

References	Participants Type	No	Participants characteristics description
[56]	Professional caregivers	7	7 female (average age = 31)
[57]	Users	13	Older adults—(age 60–80)
[58]	Users	98	Older adults—98 (age 64–78)
[59]	Users	16	Older adults (age 65+) with a risk or history of falling, recent illness, or surgery
[60]	Users	7	Older frailty patients—(average age 77, 26)—2 male, 7 female
[61]	Potential users	36	Adults—13 (<60), 14 (age 60+), and 9 AADS patients (apraxia and action disorganization syndrome)

[51], chronically ill people [46], formal professional caregivers [56], or informal caregivers [52]). In some cases, getting a sufficient number of participants from a particular stakeholder group who would be able and willing to participate in usability testing and evaluation may be challenging. In such a case, all that remains is to replace them with another group and hope that the results will not be too distorted [53].

Furthermore, even when we get participants, we must consider their limitations. For example, we often cannot carry out long and demanding testing with a few participants, but short, easy testing with more participants. Moreover, the rule of 5 participants [65] is not applicable in this case due to high intra-group variability.

Enhanced living environments are designed to improve the quality of life for individuals by providing personalized, adaptable, and intelligent services and applications in different domains such as health, well-being, entertainment, and communication. However, if these environments are not designed with usability in mind, they may not be effective in achieving their intended goals and may even cause frustration and dissatisfaction among users. Usability testing in ELE research is relatively new, evolving together with the developments in IoT technology. As IoT technology brings new options almost every day, we need to investigate how to use them effectively without barriers for all stakeholders of ELE technology.

Our study showed that usability evaluation in ELE research is mainly a European topic for different reasons. European countries have a strong tradition of social policies that aim to support the health and well-being of their citizens, the EU has provided significant funding for ELE research through various programs, and ELE research in the EU is based on strong collaboration involving multiple partners from academia, industry, and government. Expanding the research to a global level would help significantly to find appropriate methods and procedures for usability evaluation in this field.

Practitioners are still learning to apply usability engineering methods in the field of ELE and to cope with its specifics. More profound studies in optimizing these methods can only be expected in the future. The heuristic evaluation of usability in the field of ELE is hardly applied at present because heuristic criteria that correspond to this field are not defined. Over time, practitioners will gain more profound practical experience, which they transform into these heuristic rules, and therefore heuristic evaluation will be used much more often than now.

Another direction in the usability testing can be automated or remote usability testing which involves using software tools to simulate user interactions and test the usability of a product or service. Automated usability testing is faster, more efficient, and less expensive than traditional manual testing, and can be used to identify potential usability issues early in the development process. Above that, remote usability testing incorporates online tools such as video conferencing or remote screen-sharing software.

4. Conclusion

Developments in ELE research manifest researchers' efforts to discover new advanced technologies (IoT) that make staying at home and living independent life possible even for people with some limitations. This chapter showed that usability evaluation of IoT technologies in the ELE context is evolving, but it is still at its beginning. The presented literature review revealed that most studies tested the usability of ELE technology by the end users, who then filled out a questionnaire expressing their feelings about using the given technology. It also showed that none of the evaluated studies explicitly justified the choice of the usability evaluation method. Moreover, none of the monitored studies compared the performance and appropriateness of different usability engineering methods by simultaneously applying different methods to the same problem and comparing the outputs of these methods.

In the future, specific usability engineering methods must be considered: the diversity of users, the limitations of elderly and persons with disabilities, the complexity of these systems, and the rapid development of IoT technologies. ELE usability must meet the requirements of all these stakeholder groups (users, physicians, caregivers, family members and friends, and other service providers), each of which has very different characteristics. Limitations of given persons affect the ability to use given technologies, even the ability to test and evaluate their usability from the position of future users. An example could be the provision of informed consent by a participant who does not have the cognitive capacity to provide such consent. The complexity of ELE systems requires far more complex usability testing and evaluation when one does not focus only on one single part of this system but on the system as a whole. At the same time, it should be remembered that changing even a single element of this system will change the entire system and its usability. The rapid development of new technologies causes new and new tools and possibilities to appear, which are incorporated into existing systems. These specificities represent challenges that will have to be overcome.

References

[1] A.H. Bayer, L. Harper, Fixing to Stay: A National Survey of Housing and Home Modification Issues, AARP, 2000.

[2] R. Tarricone, A.D. Tsouros, Home Care in Europe: The Solid Facts, WHO Regional Office Europe, 2008.

[3] United Nations, Department of Economic and Social Affairs, World Population Ageing 2019, Highlights, 2019.

[4] OECD/WHO, Pricing Long-Term Care for Older Persons, WHO, Paris, 2021.

[5] R. Bastardo, A.I. Martins, J. Pavão, A.G. Silva, N.P. Rocha, Methodological quality of user-centered usability evaluation of ambient assisted living solutions: a systematic literature review, Int. J. Environ. Res. Public Health 18 (2021) 11507.

[6] A. Mosenia, N.K. Jha, A comprehensive study of security of internet-of-things, IEEE Trans. Emerg. Top. Comput. 5 (2016) 586–602.

[7] J.M. Paredes-Parra, A. Mateo-Aroca, G. Silvente-Niñirola, M.C. Bueso, Á. Molina-García, PV module monitoring system based on low-cost solutions: wireless raspberry application and assessment, Energies 11 (2018) 3051.

[8] V. Hassija, V. Chamola, V. Saxena, D. Jain, P. Goyal, B. Sikdar, A survey on IoT security: application areas, security threats, and solution architectures, IEEE Access 7 (2019) 82721–82743.

[9] J. Sengupta, S. Ruj, S.D. Bit, A comprehensive survey on attacks, security issues and blockchain solutions for IoT and IIoT, J. Netw. Comput. Appl. 149 (2020) 102481.

[10] L. Antao, R. Pinto, J. Reis, G. Gonçalves, Requirements for testing and validating the industrial Internet of things, in: Proceedings of 2018 IEEE International Conference on Software Testing, Verification and Validation Workshops, IEEE, 2018, pp. 110–115.

[11] M. Aazam, I. Khan, A.A. Alsaffar, E.N. Huh, Cloud of Things: Integrating Internet of Things and cloud computing and the issues involved, in: Proceedings of 2014 11th International Bhurban Conference on Applied Sciences & Technology, 2014, pp. 414–419.

[12] Cisco, The Internet of Things Reference Model, CISCO, 2014.

[13] A. Haidine, S. El Hassani, A. Aqqal, A. El Hannani, The role of communication technologies in building future smart cities, Smart Cities Technol. 1 (2016) 1–24.

[14] C.A. Díaz, et al., IoToF: a long-reach fully passive low-rate upstream PHY for IoT over fiber, Electronics 8 (2019) 359.

[15] D. Dobrilović, Networking technologies for smart cities: an overview, INDECS 16 (2018) 408–416.

[16] Z. Bakhshi, A. Balador, J. Mustafa, Industrial IoT security threats and concerns by considering Cisco and Microsoft IoT reference models, in: 2018 IEEE Wireless Communications and Networking Conference Workshops, 2018, pp. 173–178.

[17] H. Cao, M. Wachowicz, C. Renso, E. Carlini, Analytics everywhere: generating insights from the internet of things, IEEE Access 7 (2019) 71749–71769.

[18] R.K. Barik, R. Priyadarshini, H. Dubey, V. Kumar, K. Mankodiya, FogLearn: leveraging fog-based machine learning for smart system big data analytics, Int. J. Fog Comput. 1 (2018) 15–34.

[19] Y.J. Lin, C.F. Tan, C.Y. Huang, Integration of logic controller with IoT to form a manufacturing edge computing environment: a premise, Procedia Manuf. 39 (2019) 398–405.

[20] N.H. Tran, H.S. Park, Q.V. Nguyen, T.D. Hoang, Development of a smart cyber-physical manufacturing system in the industry 4.0 context, Appl. Sci. 9 (2019) 3325.

[21] G. Marques, R. Pitarma, N.M. Garcia, N. Pombo, Internet of things architectures, technologies, applications, challenges, and future directions for enhanced living environments and healthcare systems: a review, Electronics 8 (2019) 1081.

[22] Y. Fathy, P. Barnaghi, R. Tafazolli, Large-scale indexing, discovery, and ranking for the Internet of Things (IoT), ACM Comput. Surv. 51 (2018) 1–53.

[23] J. Contreras-Castillo, S. Zeadally, J.A. Guerrero Ibáñez, A seven-layered model architecture for internet of vehicles, J. Inf. Telecommun. 1 (2017) 4–22.

[24] R. Hervás, J. Fontecha, D. Ausín, F. Castanedo, D. Lopez-de-Ipina, J. Bravo, Mobile monitoring and reasoning methods to prevent cardiovascular diseases, Sensors 13 (2013) 6524–6541.

[25] D.D. Vaziri, et al., Exploring user experience and technology acceptance for a fall prevention system: results from a randomized clinical trial and a living lab, Eur Rev Aging Phys Act 13 (2016) 1–9.

[26] J. Nielsen, Usability Engineering, Morgan Kaufmann, 1994.

[27] S. Ovaska, Usability as a goal for the design of computer systems, Scand. J. Inf. Syst. 3 (1991) 47–62.

[28] J.A. McCall, Quality factors, in: Encyclopedia of Software Engineering, 2002.

[29] R.B. Grady, D.L. Caswell, Software Metrics: Establishing a Company-Wide Program, Prentice-Hall, 1987.

[30] R.B. Grady, Practical Software Metrics for Project Management and Process Improvement, Prentice-Hall, 1992.

[31] International Organization for Standardization, ISO 9126-1: Software Engineering Product quality Part 1: Quality model, 2001.

[32] International Organization for Standardization, ISO 9241-110: Ergonomics of Human-system Interaction-Pt. 110: Dialogue Principles, ISO, 2006.

[33] M. Macleod, R. Bowden, N. Bevan, I. Curson, The MUSiC performance measurement method, Behav. Inf. Technol. 16 (1997) 279–293.

[34] M.Y. Ivory, An Empirical Foundation for Automated Web Interface Evaluation, University of California, Berkeley, 2001.

[35] A.N. Badre, Shaping Web usability: interaction design in context, Ubiquity, 1, (2002).

[36] J. Nielsen, Usability inspection methods, in: Conference Companion on Human Factors in Computing Systems, 1994, pp. 413–414.

[37] C. Ghaoui (Ed.), Encyclopedia of Human Computer Interaction, IGI Global, 2005.

[38] J. Foley, W.C. Kim, S. Kovacevic, K. Murray, UIDE—an intelligent user interface design environment, in: Intelligent User Interfaces, 1991, pp. 339–384.

[39] B. Shneiderman, C. Plaisant, M.S. Cohen, S. Jacobs, N. Elmqvist, N. Diakopoulos, Designing the User Interface: Strategies for Effective Human-Computer Interaction, Pearson, 2016.

[40] A. Dix, J. Finlay, G.D. Abowd, R. Beale, Human-Computer Interaction, Pearson Education, 2004.

[41] J. Nielsen, R. Molich, Heuristic evaluation of user interfaces, in: Proceedings of the SIGCHI Conference on Human Factors in Computing Systems, 1990, pp. 249–256.

[42] J. Rieman, M. Franzke, D. Redmiles, Usability evaluation with the cognitive walkthrough, in: Conference Companion on Human Factors in Computing Systems, 1995, pp. 387–388.

[43] J. Brooke, SUS—a quick and dirty usability, in: Usability Evaluation in Industry, Taylor & Francis, 1996, pp. 189–194.

[44] M. McGee, Usability magnitude estimation, in: Proceedings of the Human Factors and Ergonomics Society Annual Meeting, 47, 2003, pp. 691–695.

[45] A. Rich, M. McGee, Expected usability magnitude estimation, in: Proceedings of the Human Factors and Ergonomics Society Annual Meeting, 48, 2004, pp. 912–916.

[46] D.R. Kaufman, V.L. Patel, C. Hilliman, P.C. Morin, J. Pevzner, R.S. Weinstock, J. Starren, Usability in the real world: assessing medical information technologies in patients' homes, J. Biomed. Inform. 36 (2003) 45–60.

[47] S. Prescher, A.K. Bourke, F. Koehler, A. Martins, H.S. Ferreira, T.B. Sousa, J. Nelson, Ubiquitous ambient assisted living solution to promote safer independent living in older adults suffering from co-morbidity, in: 2012 Annual International Conference of the IEEE Engineering in Medicine and Biology Society, 2012, pp. 5118–5121.

[48] C. Or, D. Tao, Usability study of a computer-based self-management system for older adults with chronic diseases, JMIR Res. Protoc. 1 (2012) e2184.

[49] J. Luengo-Polo, D. Conde-Caballero, B. Rivero-Jiménez, I. Ballesteros-Yáñez, C.A. Castillo-Sarmiento, L. Mariano-Juárez, Rationale and methods of evaluation for ACHO, a new virtual assistant to improve therapeutic adherence in rural elderly populations: a user-driven living lab, Int. J. Environ. Res. Public Health 18 (2021) 7904.

[50] F.M. Ligons, C. Mello-Thoms, S.M. Handler, K.M. Romagnoli, H. Hochheiser, Assessing the impact of cognitive impairment on the usability of an electronic medication delivery unit in an assisted living population, Int. J. Med. Inform. 83 (2014) 841–848.

[51] M. Kaczmarek, A. Bujnowski, K. Osiński, E. Birrer, T. Neumann, B. Teunissen, Ella4Life virtual assistant-user centered design strategy-evaluation following labolatory tests, in: 2020 13th International Conference on Human System Interaction, 2020, pp. 307–311.

[52] C.A. Byrne, M. O'Grady, R. Collier, G.M. O'Hare, An evaluation of graphical formats for the summary of activities of daily living (adls), in: Healthcare, vol. 8, 2020, p. 194.

[53] J. Soar, J. Symonds, Human interface considerations for ambient assisted living systems, J. Health Inform. 5 (2010) 1–7.

[54] G. Brajnik, C. Giachin, Using sketches and storyboards to assess impact of age difference in user experience, Int. J. Hum. Comput. 72 (2014) 552–566.

[55] I.L. Držanič, V. Mladenović, M. Debevc, V. Dolničar, A. Petrovčič, S.H. Touzery, I. Kožuh, Usability testing of a smartphone telecare application for informal caregivers, in: International Conference on Human Centered Computing, 2019, pp. 252–265.

[56] D. Bacchin, P. Pluchino, A.Z. Grippaldi, D. Mapelli, A. Spagnolli, A. Zanella, L. Gamberini, Smart co-housing for people with disabilities: a preliminary assessment of caregivers' interaction with the DOMHO system, Front. Psychol. 12 (2021) 734180.

[57] K.L.H. Ting, M. Lewkowicz, From prototype testing to field trials: the implication of senior users in the evaluation of a social application, Procedia Comput. Sci. 67 (2015) 273–282.

[58] P. Caleb-Solly, S. Dogramadzi, C.A. Huijnen, H.V.D. Heuvel, Exploiting ability for human adaptation to facilitate improved human-robot interaction and acceptance, Inf. Soc. 34 (2018) 153–165.

[59] E. Brox, S.T. Konstantinidis, G. Evertsen, User-centered design of serious games for older adults following 3 years of experience with exergames for seniors: a study design, JMIR Serious Games 5 (2017) e6254.

[60] E. Pedroli, P. Cipresso, L. Greci, S. Arlati, A. Mahroo, V. Mancuso, A. Gaggioli, A new application for the motor rehabilitation at home: structure and usability of Bal-App, IEEE Trans. Emerg. Top. Comput. Secur. 9 (2020) 1290–1300.

[61] M. Pastorino, A. Fioravanti, M.T. Arredondo, J.M. Cogollor, J. Rojo, M. Ferre, A.M. Wing, Preliminary evaluation of a personal healthcare system prototype for cognitive eRehabilitation in a living assistance domain, Sensors 14 (2014) 10213–10233.

[62] S.Z. Ahmed, A comparison of usability techniques for evaluating information retrieval system interfaces, Perform. Meas. Metr. 9 (1) (2008) 48–58.

[63] A. Doubleday, M. Ryan, M. Springett, A. Sutcliffe, A comparison of usability techniques for evaluating design, in: Proceedings of the 2nd Conference on Designing Interactive Systems: Processes, Practices, Methods, and Techniques, 1997, pp. 101–110.

[64] M.W. Jaspers, A comparison of usability methods for testing interactive health technologies: methodological aspects and empirical evidence, Int. J. Med. Inform. 78 (2009) 340–353.

[65] J. Nielsen, T.K. Landauer, A mathematical model of the finding of usability problems, in: Proceedings of the INTERACT'93 and CHI'93 Conference on Human Factors in Computing Systems, 1993, pp. 206–213.

About the authors

Dr. Hana Kopackova is an associate professor in System Engineering and Informatics at the Faculty of Economics and Administration of the University of Pardubice. Her research interest covers all the issues of digital transformation in the private and public sectors. During the last five years, she has been a member of research teams dealing with the influence of information technology on the development of smart cities and regions with an emphasis on participatory technologies. As part of her teaching activities, she is the guarantor of several subjects in Czech and English language. She is the author or co-author of more than 60 articles published in renowned professional journals or presented at international scientific conferences. Her articles are well cited, and she has an H-Index of 7 according to the latest Scopus sources. She is also the author or co-author of four monographs and four university textbooks. In 2020 and 2021, she worked as deputy head at the Institute of System Engineering and Informatics. Since 2022, she has been appointed as vice-dean for internal affairs.

Dr. Miloslav Hub is an Associate Professor in the Institute of Institute of System Engineering and Informatics at Faculty of Economics and Administration University of Pardubice. He is an expert in the field of Usability Engineering. He has published many international journal articles in the broad areas of Computer Science Engineering, Software Modeling and Testing, Software Security, Usability Engineering, Intelligent Systems etc. His articles are well cited, and he has an H-Index of 8 according to the latest Scopus sources. His major contributions are in Usability Engineering, data Security, Biometric Authentication and Artificial Intelligence. In his decade-long academic career, he was invited as the keynote speaker in many reputed conferences and has also delivered many guest lectures abroad. He is always passionate and enthusiastic in updating himself on the recent trends in technology and education. In addition, he co-authored a book titled "Selected Aspects of Cyber Security" and has also published several book chapters with reputed publishers.

CHAPTER SIX

Internet of Things to enhanced living and care environments for elderly: Applications and challenges

Analúcia Schiaffino Morales[a] (iD), **Ione Jayce Ceola Schneider[b]** (iD), **Fabrício de Oliveira Ourique[a]** (iD), **and Silvio César Cazella[c]** (iD)

[a]Federal University of Santa Catarina – Sciences, Technologies and Health Education Center – Computer Department and Graduate Program in Energy and Sustainability, Santa Catarina, Brazil
[b]Federal University of Santa Catarina – Sciences, Technologies and Health Education Center – Health Science Department, Santa Catarina, Brazil
[c]Federal University of Health Sciences of Porto Alegre – Department of Exact Sciences and Applied Social – Information Technology and Healthcare Management, Rio Grande do Sul, Brazil

Contents

Abstract

Aging is a life's phase that presents challenges due to lower functional capacity that can be a consequence of poor quality of life. Technologies such as the Internet of Things, have helped to improve the quality of life for elders. Applications in intelligent environments such as hospitals, are not a reality for most countries, but once incorporated they allow for improved quality of life due to the autonomy and independence

Advances in Computers, Volume 133
ISSN 0065-2458
https://doi.org/10.1016/bs.adcom.2023.10.005

153

provided. This chapter presents a rapid literature review of advances including the Internet of Things and Living Environment related to elderly and healthy aging based on academic bases: PubMed, Scopus, and IEEE Xplorer. The PRISMA was applied, and 58 papers were classified into 10 categories, including design and investigative approaches. Challenges related to systems of care for the aging were discussed and it was concluded that questions about the importance of senior adaptability, privacy and security issues deserve further attention.

1. Introduction

Aging process is a social achievement and a challenge, too. The world expects to have 1.2 billion people over 60 by 2025, and that number is expected to exceed 2 billion by 2050 [1,2]. The form people and society think and act concerning aging process are extremely important to improve this phase of the life cycle [2]. Politically, the aging process is often seen as onerous, because it is synonymous with a sick person, dependent on care, who lives in social isolation, with many deprivations and physical limitations [3,4]. However, it can be a phase in which the person remains the protagonist of his life even with limitations imposed by the process and due to some disease [3,5].

In the 1990s, World Health Organization adopted the concept of "active aging." Based on this concept, the person is expected to recognize their physical and mental well-being. In addition, people are encouraged to participate in society according to their needs and abilities, in environments with adequate security and assistance. With this, people have an increase in healthy life expectancy [1]. This concept seeks to demystify that the older adults are synonymous with retirement, illness and care dependent [1]. Healthy aging is being able to develop and maintain functional capacity for well-being. This includes two important concepts: intrinsic capacity, which includes physical and mental capacity, and functional capacity, combining environment, individual, and health-related attributes [6].

Functional assessment of the older adult should be a pillar for decision-making on adequate diagnosis, prognosis, and treatment. It allows assessing at what level certain diseases or conditions prevent autonomous and independent performance of activities of daily living [7]. Lower functional capacity demands more complex health care. Public health policies for the elderly should prioritize the maintenance of functional capacity; independence in their routine activities; autonomy, and participation in their own care. Thus, encouraging prevention and comprehensive health care [5].

The International Classification of Functioning, Disability and Health (ICF) reports that functionality is the result of physical, environmental and social conditions [8]. Performance during activities may vary depending on the environment, and the individual. A person with a lack of partial integrality in physical conditions will be able to perform their functions if the environment is adapted to their limitations. Likewise, society may impose barriers and limit an individual's functions through judgment and lack of acceptance [9]. Comprehensively, the term disability is used when the elderly person has some disability, limitations and restrictions of activities [10]. There are several causes of disability in performing daily activities, such as senile dementia, cognitive impairment, Alzheimer's disease, obesity, sarcopenia, head trauma, low back pain, and others [11–17].

The loss of functional abilities does not occur only due to the underlying disease, but also due to cognitive impairment and physical capacity [18,19]. The first skills to be lost are the most complex skills that require the person's ability to deal with themselves and social environments [18–23], followed by those exercised in their daily lives [20,22,24]. However, as the cognitive impairment worsens, there is a decrease in the abilities for daily activities [19]. This results in the reduction of quality of life [25], with implications for the individual, family and society [23] and increased costs to health systems [19].

The accumulation of chronic diseases in life is known as multimorbidity. This condition worsens functional limitations. In addition, mood disorders are related to multimorbidity and worsening of functional limitations [26,27]. Another condition that impacts functional limitations is obesity. Furthermore, the mechanisms of obesity lead to an increased risk of multimorbidity [28]. The coexistence of chronic and mental conditions in functional limitations has an impact on health policies and clinical practice [27].

Enhanced Living Environments (ELE) solutions demand suitable algorithms, architectures, and platforms, considering the advance in technologies in this area and the development of new and innovative connected keys such as cloud computing, fog, and edge solutions, advance in communication protocols such as Bluetooth Low Energy (BLE), 6lowPAN, WIFI and the 5G networks. Then this area has the most challenges to resolve such as social concerns, adaptability attributes, and issues about privacy and security still without a market good solution.

There are many efforts to develop solutions including smart homes assisted for chronic disease patients and the elderly population. In the meanwhile, several solutions have been published in the scientific literature adopting the Internet of Things (IoT). The new paradigm is based on several objects interconnected collecting data about things and reporting to a cloud computing system to improve daily activities. Emphasizing a particular point about the conditions in the residences and buildings which degraded the health of the aging population about the indoor air-quality monitoring because it is normal for these people to stay much time in the interior of their residences [29]. Real-time air-quality monitoring must avoid a bad situation if some kind of chemical particle or another toxic substance category appears in the aging people's homes, for example [29,30]. On another hand, offering health services considering the advances in people aging and the recent COVID-19 outcomes will increase the costs and investments of the government in hospitals, health clinics, health workers, and nursing care to attend to the increasing demand. The IoT may be an alternative to improve the aging population to save their quality of life. In this sense, telemedicine and remote caregivers must complement the needs of the aging people through communications distance services and home assistance possibilities [29–32].

It is uncommon to find works reporting applied results from pervasive health systems, which isn't reporting a prototype experience [33]. Therefore, this chapter investigates the ELE applications in the IoT domain and attempts to determine what categories of applications are already experienced by aging people volunteers or are found in operation. For this purpose, a review is presented and described in Section 2. The forecasted outcomes are categorized through the main goal identified during the review process. They are home care assistance systems, monitoring of daily living activities, air quality supervision and buildings, fall detection, chronic illnesses monitoring, indoor localization, and other applications which concatenate three individual findings. Section 3 presents the results and a brief analysis from the viewpoint of the proposed methodological questions. The discussion is provided in Section 4, following the chapter Summary and References.

2. Research methodology

Developing a responsive, timely and credible, and respecting critical principles of knowledge synthesis [34] were investigated in three databases to encourage a rapid review of the roadmap of the advancements, including the Internet of Things and Enhanced Living Environment related to elderly and aging healthy. This study adopted the Preferred Reporting Items for Systematic Reviews and Meta-Analyses (PRISMA) recommendations [35] and the Rayyan tool [36]. The search was conducted through an adapted PIE strategy [37], changing patients for the population of interest (Population, Intervention, Effect/Assessment) as the same scheme established in Ref. [38].

2.1 Search process and research questions

Following the components of the PIE strategy, search terms and databases investigated:

- Population: elderly, older and aging people
- Intervention: Internet of Things, IoT and enhanced (assisted) living environments (ambient)
- Effect/Assessment: applications, technologies, strategies
- Search string applied: (enhanced OR assisted) AND living AND (environments OR ambient) AND internet of things.
- Databases: Scopus, PubMed and IEEE Xplorer: Scopus, 603, PubMed, 90 and IEEE Xplorer, 58.

The general questions guide the rapid review follows.

- GQ1: How can the Internet of Things help health systems with population aging?
- GQ2: What applications and resources are already being tested in the elderly people or are operationally ready?
- GQ3: Has any evaluation study about IoT and assisted environments for elderly people?
- GQ4: Have any of these new strategies already been tested and adopted or are they being commercialized?

The strategy for the study was to find words to rank and help the selection of the papers, through the ranking resources provided by Rayyan that extract all the words in the documents and count them as a table shown in bar graph in Fig. 1.

2.2 Literature search and article selection

Table 1 describes the inclusion and exclusion criteria applied in the study. We had an objective to investigate through a rapid review and some exclusions needed to be considered about the conference paper because this kind of research paper is very short to help in mapping desired. Also, we choose to drop reviews and book chapters at this moment.

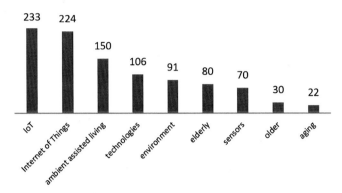

Fig. 1 Ranking keywords for inclusion election.

Table 1 Inclusion and exclusion criteria.

Inclusion criteria	Exclusion criteria
Articles published in international journals are written in English	Duplicate articles
Articles with implementation results or evaluation outcomes	Reviews papers: systematic reviews, systematic mappings, surveys, any kind of review
Articles available in their full version	Articles that do not allow open access
	Not qualifying as an article, although being classified as such in a journal (editorials, book reviews, etc.)
	Conference papers and book chapters

Automatically, the Rayyan tool identified 150 duplicate works. It was confirmed manually and deleted 75 papers. In the screening phase, it was applied excluded rules, first, found the primary research about the ELE and IoT scenario, then excluded different kinds of papers that were not eligible for this study, such as conference papers, books, chapter books, reviews, surveys, systematic reviews, they were counting as "wrong publication type." In total 431 papers were excluded by the title and type of paper. Additionally, after reading the abstracts were reviewed, and added 156 papers to the exclusion criteria. This rapid review only included papers in the English language. The PRISMA flow chart shown in Fig. 2 presents

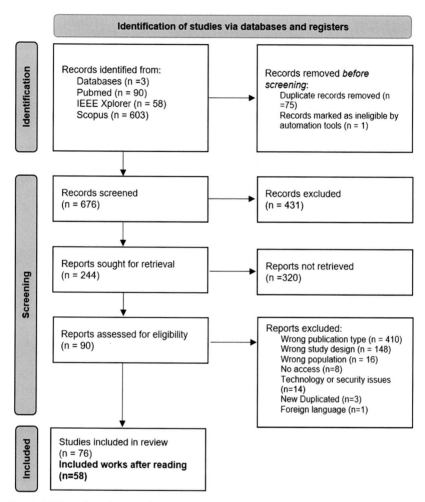

Fig. 2 PRISMA flow diagram for rapid review.

the resume of the rapid review methodology with all exclusions issues adopted and counted. There were included periodic articles with permission valid by the "Portal CAPES" subscription (labeled as "No access"). And we also extracted some papers specific for the computer science area with issues about develop security algorithms, blockchain, signal, and sound processing outcomes that were not framed in our concerns (labeled as "Technology and security specifics"). So, after the screening, we read 90 papers. There were discarded eight works that asked to pay for full work access. Finally, after the exclusion criteria were applied, 76 papers matched the eligibility criteria adopted in the study. After the reading stage, additionally, 16 papers had been excluded according to the exclusion criteria previously mentioned in Table 1.

2.3 Data extraction and synthesis

The initial data extraction is applied to all qualified publications, and the following information was obtained:

- Titles and abstract;
- Authors names;
- Publication year;
- Keywords;
- Application categorization;
- Sensors and communication technologies or machine learning techniques;
- Experimental studies with volunteers.

2.4 Risk of bias

The results from the databases were not limited to a period that reduced the bias risk. Meanwhile, there is a great risk of bias which is a limitation of this research study. Additionally, with the subjectivity of inclusion and exclusion criteria employed by the authors in the screening, we investigated only three databases for the data collected. Although they include great reputable works in academic research, there is an extensive range from other databases that should be considered.

3. Results

The qualified works were preliminary categorized into similar topics and findings after the abstract readings round. Table 2 presents categorized papers with labels and authors. Also, the mapping about what general question each category helps to answer also has been included as a third column.

Fig. 3 shows the distribution per year from the selected papers. There is a significant increase in concentrated publications from 2019 forward.

3.1 Answer GQ1

After reading, the main goal of each work helped to create topics representing the different application areas. Ten topics were identified: design and investigative approaches, home care assistance system, monitoring of

Table 2 Categorized findings after the readings and respective references.

Category label or suggested topic	References	Questions
Design and investigative approaches	[39–45]	GQ1, GQ2, GQ3
Home care assistance system	[46–50]	GQ1, GQ2
Monitoring of daily living activities	[51–68]	GQ1, GQ2
Air quality supervision and buildings	[69–76]	GQ1, GQ2
Fall detection system	[77–81]	GQ1, GQ2
Chronic illnesses	[82–84]	GQ1, GQ2
Robotic systems	[85–87]	GQ4
Solutions available and evaluation	[88–90]	GQ3
Indoor localization	[91–93]	GQ1, GQ2
Others works	[94–96]	GQ1, GQ2

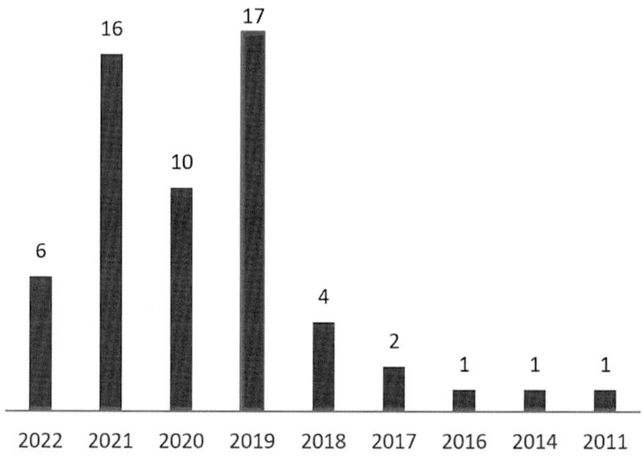

Fig. 3 Distribution of selected papers for publication year starting by more recent papers.

daily living activities, air quality supervision and buildings, fall detection system, chronic illnesses monitoring, solutions available and evaluation, indoor localization, and other applications (see Table 2). Fig. 4 shows the graph with proportion of findings in the rapid review and the suggest categorized areas.

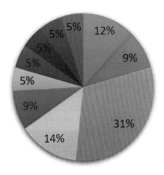

- design and investigative approaches
- monitoring of daily activities
- fall detection system
- robotic systems
- indoor localization
- home care assistance system
- air quality and buildings
- chronic illnesses
- solutions avaliable and evaluation
- others applications

Fig. 4 Topics distribution of rapid review selected papers.

3.1.1 Design and investigative approaches

Different articles about challenges concern the elderly people and their quality of life, as well as several studies including how to design and evaluate the IoT-based systems, were grouped in this category. Articles such as geriatric mapping problems involving assisted daily living were identified through questionnaires applied in nine centers in Malaysia. Where the findings established the main challenges for the technology application should include Dementia, Emotional Instability, and Chronic Degeneration [39]. Although these results are for care centers the most significant challenges faced by the elderly in this context might be used for engaging researchers' concerns with other kinds of applications demanded to enhance the living environment for the aging people in their residences. Another work investigating the dementia, which is the degeneration of the cerebral cortex, the part of the brain responsible for thoughts, memories, actions, and personality, usually causes dementia were modeled using a CASAS dataset and machine learning techniques. Designing to collect real-life data through a set of sensors including environmental information, this study keeps the caregivers informed and alerted them about abnormal situations. They designed a scheme including several weeks of observation of elderly households in a region of Italy. They were several objects and environments with different kinds of sensors collecting the elderly living situation which should be confirmed in Ref. [40]. Another work investigating dementia, which is the degeneration of the part of the brain responsible for thoughts, memories, actions, and personality, which usually triggered dementia was modeled using a CASAS dataset, recording a total of 400 participants who performed the daily activities and the employment of machine learning techniques. The outcomes were published in Ref. [44]. All works grouped in this category present instigating experiences with elderly volunteers consisting of valuable information for the person interested in this research domain Refs. [41–43].

3.1.2 Home care assistance system
In this category, there were six published papers between 2017 and 2019, that have differences from the recently published articles, such as concerns about monitoring daily activities. A Wireless Acoustic Sensor Network is designed for AAL systems processing the environmental sounds and rapidly able to identify situations, as a audio fingerprint recorded in Ref. [46]. The Habitat project case study has been mentioned in two Refs. [48,49]. Emphasizing the home assistance importance to the autonomy of more aged people and the usage of the oldest technology such as RFID tags for object identification such as smart chairs to identify the incorrect postural positions, or wearable belt to identify the indoor space movements. Most of the solutions were prototypes, and just one case was tested within pilot tests, in Ref. [50] they tested with older people living alone driving the focus on the development process. This work also employed classification methods for the pattern recognition denominated a Smart-Habits which is an intelligent privacy-aware system for persons living alone, identifying unusual situations. Also, in this set of selected papers, a low-cost proposal capable of recording ambient sounds based on audio capture provides a different approach to the sensor network.

3.1.3 Monitoring of daily living activities
The rapid review method identified 18 papers with systems or prototypes for monitoring daily living activities. Several works in this review were ranked as laboratorial proof of concept set, considering these groups about older people living activities specifically, were identified 10 papers which achieved experimental tests with volunteers' elderly people. The investigation of this category of papers concentrates on employing different kind of sensors to gather information about the daily-living activities and transmit it to a cloud platform, and then synchronize with caregivers or other identified people. Three of them invested in the dementia assistance in their studies

mechanisms to helping older adults with advanced cognitive impairments, or mental diseases, such as in Refs. [44, 68, 97]. A mentioned gap [56] is considered the individual requirements assistance and it is characterized by a gradual cognitive and physical decline from aging. Several issues about health care are so personal and individual, and the challenge to develop unique solutions should be adapted to the elderly people's features. For example, different biomarkers for physiological health conditions indicate an individual routine, cognitive debt is very common in the aging process, and the different chronic illnesses may be associated with each person. In general, the work concerns non-invasive monitoring and keeping the care-givers up to date about conditions through the IoT sensors and platforms. The efforts compound substantial significance to avoiding elderly hospital-ization and decreasing the socio-economic costs of caring, while it may also significantly enhance senior people's quality of life. The solutions presented in these works employed Bluetooth Low Energy (BLE) for communication in eight matched cases. The second protocol most encountered in the arti-cles was WIFI IEEE 802.11 distribution in their solution. There were no references to security or privacy, remembering that we excluded the specific works with security and privacy investigations in this selection method. About the systems functionalities, tracking or indoor navigation, memory training, user interactions, diagnostic, and general diary activities were the most frequent. Some articles recorded the preoccupation with the elderly adaption to technology. This is most evident with a solution that concerns ontologies and voice recognition commands, such as the case presenting a voice assistant application for avoiding sedentary in elderly people, one of the tested applications with volunteers between 70 and 90 years old reported in Ref. [60]. The voice assistant is alternative for technology usage consid-ering that most aging people have difficulty seeing or handling technological equipment. This interface choice should be considered in several other circumstances to improve the aging people's experience with technological apparatus. Also, it should include emotional detection or a similar through the wearables or cameras, or any kind of sensor already used for other explo-rations in the solutions identified. This information is very important to evaluate the elderly condition of the proposed IoT system. Finally, six papers employed the machine learning classification algorithms in their solutions Refs. [44, 63, 68, 97]. And deep learning and LSTM were applied to identify long-term personalized monitoring of the daily activity. This work is dedicated to inferring abnormal behaviors, such as unhealthy or emergent cases [65]. And one paper with Radial Basis Function Neural Network and auto encoder modeling solution is presented in Ref. [55]. This paper used four binary datasets registering daily living activities. The proposed method is a hybrid artificial neural network to recognize home-based activities.

3.1.4 Air quality supervision and buildings
Indoor buildings' air monitoring is an essential application, because of the majority part of people's lives have been spent in their homes [29,49]. For healthy aging and enhanced living for the future of this population, it is necessary to increase the investigations into the quality air domain. From the findings of the rapid review methodology, only one author has excelled in this domain, presenting several works and leading international research groups to demonstrate the importance of monitoring indoor air quality in the advancement of enhancement of quality of life. One of the works presents an open-source proposal that operates with Wi-Fi and has been a cost-effective alternative to deploying this kind of solution [69]. Different studies were very well explored in works published by this group. Emphasizing the IoT-based solutions and indoor air quality possibilities to be deeply investigated in the next few years. Their research has also explored the concentrations' levels for measurements sampled through prototypes and practical applications [70,72–74]. Another interesting approach finding is based on machine learning classification to recognize human activity. The work presents an IoT air quality sensor dataset and an architecture with hardware and software components collecting the data in real time. They present satisfactory results for the three kinds of classification process proposed (normal situation, smoke presence, and single activity), evaluating the success through machine learning accuracy [75]. A method of prediction employing physiological parameters identified with sensors measuring the hand skin temperature and pulse rate, along with the ambient air temperature. According to the presented study, people spend approximately 90% of their daily time in indoor spaces justifying the importance of assure thermic comfort and healthy indoor environments. With the increase of aging people, this kind of study is essential to improve the habitat and social conditions in the building's domains [76].

3.1.5 Fall detection systems
Fall detection is an important issue in the elderly IoT application context. Maybe one of the most dangers to older people living alone, due the age advance there is several disorders contributing to the possibility or probability to fall indoors mainly. For example, rheumatics and musculoskeletal

disturbing, mental confusion, or even a simple distraction. All recent works were published between 2020 and 2022, and all studies providing artificial intelligence-based solutions include several datasets in their investigations. This kind of application has been encountered in modern smartwatches, meanwhile, these selected papers report innovative solutions applying the sensor information from datasets directing to the concern of falls and their severity. All papers tested and implemented different machine learning for the fall detection application. Convolutional neural networks for feature extraction associated with eXtreme Gradient Boosting classification resulting in 88% of accuracy were reported by Ref. [77]. In Ref. [80] a fall detection framework uses deep learning and 5G networks including an edge computing proposal to reduce the cost of communication infrastructure and reduce the latency and consequently the bandwidth for the proposed solution. Presenting a different approach Ref. [78], tested several different machine learning algorithms through the Rapid Miner tool. The proposed CLeFAR approach in Ref. [79] combines the traditional classification scheme fuzzy argumentation approach identifying the activity behind the occurrence of falls. And finally, an architecture based on fog-cloud 3-layer deploys deep learning reporting a high potential to improve the accuracy of fall detection applications [81].

3.1.6 Chronic illnesses

Considering the aging people or elderly as a population and all the restrictions mentioned in the rapid review methodology, resting in this category group three papers. These studies were destined to monitor elderly patients suffering from chronic diseases. There is one paper dated 2011 presenting a personal device for diabetes therapy management that uses self-monitoring blood glucose (SMBG) measurements through IoT-based AAL to help with insulin therapy [82]. Another study exploited the MIMIC-II database from Physionet containing thousands of multiparameter recordings from ICU patients between the years 2011 and 2019. Furthermore, they created a synthetic database to complement the investigation of the machine learning techniques involved. They applied a naïve Bayes–firefly algorithm (HAAL-NBFA) to monitor the chronic illness [83]. And the third work regarded on this topic is a proposed hybrid context-aware model for patients under supervision at home. The authors present a hybrid architecture with local and cloud-based elements, preventing troubles with a lost connection with the Internet. There were utilized techniques to monitor physiological signals, ambient occurrences, and activities of the patient to identify their real-time condition. According to the report, they proposed a model with time access reduced and was more accurate than other remote patient monitoring systems. It was registered as a case study of three patients monitored with blood-pressure disorders [84].

3.1.7 Robotics systems

The advanced robotics that can provide physical assistance to elderly people might be seen as a social component to integrate the issues about introducing the technology in diary living of more ancient individuals. Emphasizing this point, three studies were selected in this classification. With the aim to improve the AAL and IoT-driven environments, their approach has been employing the Microsoft Kinect and Smartwatches to increase a database destined to improve the research with assistive activities related to different postures. Their outcomes enable a robot to understand the activities of a human a sitting posture, for example, at the dining table, and then design the assistive assignments consequently [85]. The concept of the Internet of Robotic Things was introduced considering the robots as component of a IoT ecosystem. A case study supplying an implementation of three proposed patterns in the scope of work: namely, obstacle avoidance, indoor localization, and inertial monitoring was proposed and discussed in their investigation [86]. With the goal to develop a home environment for all people, including the aged and individuals with disabilities the paper proposed three robotics systems for assistive purposes considering the scope of the RSH (Robotic Smart Home) project. The first area is the mobility and transfer assist system which develops three types of systems: a lifting type, lateral-transfer type, and suspension type, all can be used for different levels of disabilities severity. The second area is denominated operational assistance combining a hand robot and environmental control system. The control system integrates IoT connections for simple home tasks such as turning on the TV, switching on lights, and air-conditioning control. And the robotic hand is destined for non-typical tasks. The third is the information assist system that is able to promote good health in daily activities using physiological measuring from sensors, and communications devices all integrated with a remote medical institution and other functionalities described in their work [87].

3.1.8 Solutions available and evaluation

We have classified three papers in this category representing the studies about evaluating applications or similar questions with real situations about commercial answers. The first, recently published (2022) presents 15 smart living environments emphasizing important points about technology adoption by aging people. In a brief analysis, six of the solution were pilots, and are now inactive; five are in the development stage; only three are currently operational and active and one was canceled. Fortunately, 13 have health as a priority, and 7 have actively assisted living considerations in their components, but only one solution was pointed as ready operationally. Additionally, some important aspects were discussed as large gaps: user acceptance, security and privacy, accessibility, and finally interoperability [88]. The second paper suggests the association of Mobile Cloud Computing and the Internet of Things to supply interconnected infrastructures that keep people to be continuously connected with other people pervasively. A prototype cloud-enabled mobile application named IPSOS Assistant was developed to watch people's well-being and handle emergencies into indoor domains [89]. And a third paper mentioning the increase of aging people in South Korea and the main challenges to integrating smart homes with IoT destined for elderly assistance. As a result, they provided an analysis about residential space and functionalities of AAL system integrating IoT resources. As a result, they provided an extensive analysis of residential space and functionalities of the AAL system integrating IoT resources. For example, the bedroom might incorporate AAL planning have a kind of motion bed control with recognition of posture or pattern for encouraging deep sleep, or a device as a sleep checker for healthcare monitoring, or still environment control devices to monitor air quality, temperature or any kind of environment information [90].

3.1.9 Indoor localization

Research about indoor localization were observed in three different papers, which exploring the transmission technology features to tracking and localization for assisted living exploiting the signal transmitted from a mobile dispositive, or even considering the signal transmitting intensity by antennas. Experimental results with radio-based indoor localization were proposed in Ref. [91] exploring the 5G transmission capabilities. And, two other findings based their experiments with the Bluetooth Low Energy (BLE) seen in Refs. [92, 93]. There are concern about the localization accuracy though BLE functionalities that provides a signal around 100 m and might not be sufficient for detection emergency situations with older people behavior. This question will be examined in the appropriated discussion section.

3.1.10 Other applications

Finally, three different findings were categorized into another applications topic. First, the advance of age, some people can suffer from musculoskeletal disorders and other illnesses already mentioned during the introduction section. Sarcopenia corresponds to decreased muscle mass, reduced strength, and impaired physical routine. The process starts with a slow rhythm between age 40 and 50 with evolution after 60 years old. The proposed system architecture with three subsystems was presented in Ref. [94] for monitoring people which suffering from this disease. Employing low-cost sensors connected on board was testbed measuring muscle parameters such as speed, muscle tone, and force. The collected information is transmitted through Bluetooth and accessed using a web application. A proof-of-concept validation was described in the paper. Considered two important issues for the safety of patients: drug compliance and adverse drug reactions are the scope of the work in Ref. [95], with an IoT approach for drug identification and monitoring. With the goal of examining drugs in order to complete treatment, to detect harmful side effects, allergies, liver or renal contradictions, and harmful side effects during pregnancy. The paper presents a technical study describing the different communication technologies considering lifetime battery, distance, and different IoT capabilities technical issues as a mechanism to help with drug administration.

3.2 Answer GQ2

The selected papers observed are in the majority of proof-of-concept systems. There are few works tested with older people volunteers, remaining several gaps in how these technologies can get social acceptance by the aging people. There is a set of findings presenting real needs from questionnaire approaches in elderly care centers. These papers have the richest material for new ideas to create IoT-based applications implicating an enormous impact on the independence and enhanced living of the senior population. The reported works already tested with volunteers in most of the cases belongs to the *monitoring of daily living activities* topic.

3.3 Answer GQ3

There is a certain lack of data regarding technology estimation or evaluation mechanisms for how the app has been performed or adapted for elderly people with IoT-based systems. Through the rapid review approach, only one work (Ref. [88]) was identified and presented an evaluation addressed to IoT and AAL domain and pointed out large gaps related to this topic. The authors discussed the research priorities such as the elderly people's adoption, security, and privacy to continue advancing AAL technologies. Specifically, the concerns about end-user testing to understand the adaptability issues have been reduced in several projects examined.

3.4 Answer GQ4

In general, the work solutions found include the following steps: acquire different signals by sensors or another kind of dispositive, then successfully transmit the data to intermediary IoT layers or to the cloud computing directly and distribute the information to caregivers or medical remote unities. These steps conceive the IoT functionality for the healthcare domain. The solutions suggest new technologies and platforms environments to supply the infrastructure for applications in this area, most have very similar stages, and few of these works have been effectively tested with the elderly population. Accordingly, with the evaluation concerns already mentioned, it is occasional to find a project with the identification of user needs as planning in an initial stage. There are few works really tested out of laboratories and a fewer number of proposed systems tested, approximately eight reported tests with volunteers. We do not find evidence about commercial usages. Another important point concerns the user acceptance, discussed in a unique paper [88], which describes some solutions for smart home systems. The paper about evaluation highlights the great challenges including privacy and security, added to interoperability of smart cities and social acceptance.

4. Discussion

Through the outcomes, it was observed significant findings mapping in this chapter. Ten topics were listed with research distributed in the strongest areas presented by each work. The works that fit the rapid review definitions were published between 2011 and 2022. And there are many challenges to attend the need of the elderly population considering the issues pointed out in the introduction, previewing a considerable and essential need for new studies and research into this domain. Most recent works incorporate a greater number of functionalities, and practically all of them offer a structure to satisfy the need of the IoT-based systems including at least three to four layers to collect the information, transmit and then distribute it to the users, caregivers, and other responsible individuals. In this study, there were found several works using BLE and WIFI to communicate, mainly the wearable or IoT sensor to acquire the information. In fact, neither all articles mentioned the type of communication protocol specification.

The topic "Monitoring of daily activities" concentrated a great number of studies, with more than 50% published between the years 2021 and 2022. This might indicate advancement for research interest in improving systems aimed to monitor the daily activities of the elderly because all studies have mentioned the concern with the increasing aging population preview. Furthermore, there is a need to investigate new approaches to help for mitigating this problem. Accordingly, in this category were found most articles with outcomes from volunteer tests, in this case, 10 studies. The lack of articles aimed at integration, or evaluation of studies from volunteers, is evident. Most of the research points to results with prototypes and some reports of proof-of-concept cases. Several solutions utilized datasets and videos, mainly proposals associated with machine learning or deep learning approaches. Some article authors accuse the COVID-19 pandemic of the absence of test investigations with elderly people.

The results of the topic "Design and Evaluation Study " concentrate on studies for planning new applications. There are some articles reporting the application of questionnaires at elderly care centers. Which can be an interesting source for anyone interested in starting or improving research in these areas. In general, all selected studies present a consistent scope of projects and correlated works. No commercial products were observed. Most prototypes seen in the articles have no concerns with their final arrangements. As well, this observation may be related to the absence of social challenges associated with the elderly. We concluded that based on the analysis of the several difficulties of aging presented in this work, this research area requires further exploration, emphasizing the needs of senior adaptability, product usability with regard to functional disabilities, as well as simple recharging of sensor batteries, for example. Technology applications and IoT-based systems must encourage healthy aging with independence and life quality. The social challenges also were registered in several articles, but no important issues were documented in the rapid review context, indicating a limitation on this point.

Finally, the machine learning works concentrates on analyses of the information collected from sensors or datasets, no works mentioned modern artificial intelligence approaches turning the proposed system explicable or emotional [98]. This approach would be applicable to improve the adaptability or to include social remarks in their investigations. Just one work destined to use voice assistance to facilitate information access and promote a better experience with the proposed technology [60].

5. Conclusion

Technology solutions can be incorporated into ELE to improve care and health for older people. Studies examined in this chapter have used different types of sensors, microcontrollers, wireless communication technologies, and software platforms. ELE systems intended for the elderly population monitor human physiological and environmental parameters using wireless communication technologies. This is to allow a certain degree of independence for individuals and to assist in medical treatments. However, there are a lot of new findings to investigate in the next years. It is evident the need to create solutions that can be adapted to the elderly, and challenges, and difficulties related to aging were pointed out. Even with people who have more active aging, there is a concern to maintain the quality of life and allow people to face old age with adequate conditions. Technology strongly presents conditions to improve assisted living environments. The main challenge to be improved is the absence of security and privacy of information in the solutions provided by the studies. It is crucial to incorporate strong security methods in order to guarantee the data's reliability, and legislation and regulations must be put in place in order to protect the rights of the users. Maybe this issue is responsible for the insufficient progress of the projects in this sector. Most reported the testbeds and proof of concept works, few demonstrate tests with volunteers and practically there are no commercial uses reported. Another point, not observed in works, is corresponding to the use of mechanisms to avoid the lack of connectivity. Solutions might consider using different communication protocols to ensure that the system is not disconnected. It is a relevant problem in an assisted environment for aging people if it was unplugged or disconnected in an emergency. To improve assisted living environments for the elderly population, these issues must be taken into account.

References

[1] World Health Organization, (2022) Noncommunicable Diseases and Mental Health Cluster Noncommunicable Disease Prevention and Health Promotion Department, and Ageing and Life Course, "ACTIVE AGEING: A POLICY FRAMEWORK Active Ageing," 2002, Accessed: Sep. 30, 2022. [Online]. Available: http://www.who.int/hpr/.
[2] World Health Organization, Decade of Healthy Ageing, WHO, Geneva, 2021.

[3] C.E. Alvares, Coimbra Junior, M.C. de Sousa Minayo, Antropologia, Saúde e Envelhecimento, Editora Fio-cruz, Rio de Janeiro, RJ, Brasil, 2002.

[4] C.F.R. Dardengo, S.C.T. Mafra, Os conceitos de velhice e envelhecimento ao longo do tempo: contradição ou adaptação? Rev. Ciências Humanas 18 (2) (2022). Oct. 2019, Accessed: Sep. 24, 2022. [Online]. Available: https://periodicos.ufv.br/RCH/article/view/8923.

[5] R. Veras, R. São, F. Xavier, Envelhecimento populacional contemporâneo: demandas, desafi os e inovações population aging today: demands, challenges and innovations RESUMO, Rev Saúde Pública 43 (3) (2009) 548–554.

[6] World Health Organization, World Report on Ageing and Health, 2015. https://apps.who.int/iris/handle/10665/186463. accessed Sep. 30, 2022.

[7] Ministério da saúde Brasília-DF, Envelhecimento e Saúde da Pessoa Idosa, 2007, Accessed: Sep. 30, 2022. [Online]. Available: https://bvsms.saude.gov.br/bvs/publicacoes/abcad19.pdf.

[8] E.S. Araujo, C.M. Buchalla, O uso da classificação internacional de funcionalidade, incapacidade e saúde em inquéritos de saúde: uma reflexão sobre limites e possibilidades, Rev. Bras. Epidemiol. 18 (3) (2015) 720–724, https://doi.org/10.1590/1980-5497201500030017.

[9] World Health Organization, World Report on Disability, 2011, Accessed: Sep. 30, 2022. [Online]. Available https://apps.who.int/iris/handle/10665/44575.

[10] C. Silveira, et al., Adaptação transcultural da Escala de Avaliação de Incapacidades da Organização Mundial de Saúde (WHODAS 2.0) para o português, Rev. Assoc. Med. Bras. 59 (3) (2013) 234–240, https://doi.org/10.1016/J.RAMB.2012.11.005.

[11] C. De Liao, J.Y. Tsauo, S.W. Huang, J.W. Ku, D.J. Hsiao, T.H. Liou, Effects of elastic band exercise on lean mass and physical capacity in older women with sarcopenic obesity: a randomized controlled trial, Sci. Rep. 8 (1) (2018), https://doi.org/10.1038/S41598-018-20677-7.

[12] S. Michalon, J.P. Serveaux, P. Allain, Frontal functions and activities of daily living in Alzheimer's disease, Geriatr. Psychol. Neuropsychiatr. Vieil. 16 (3) (2018) 321–328, https://doi.org/10.1684/PNV.2018.0749.

[13] M. Kamiya, T. Sakurai, N. Ogama, Y. Maki, K. Toba, Factors associated with increased caregivers' burden in several cognitive stages of Alzheimer's disease, Geriatr. Gerontol. Int. 14 (Suppl 2) (2014) 45–55, https://doi.org/10.1111/GGI.12260.

[14] K.K. Zakzanis, K.M. Grimes, S. Uzzaman, M.A. Schmuckler, Prospection and its relationship to instrumental activities of daily living in patients with mild traumatic brain injury with cognitive impairment, Brain Inj. 30 (8) (2016) 986–992, https://doi.org/10.3109/02699052.2016.1147077.

[15] C.D.F. de Souza, et al., Physical disability degree in the elderly population affected by leprosy in the state of Bahia, Brazil, Acta Fisiátrica 24 (1) (2017), https://doi.org/10.5935/0104-7795.20170006.

[16] A. Lardon, J.D. Dubois, V. Cantin, M. Piché, M. Descarreaux, Predictors of disability and absenteeism in workers with non-specific low back pain: a longitudinal 15-month study, Appl. Ergon. 68 (2018) 176–185, https://doi.org/10.1016/J.APERGO.2017.11.011.

[17] P.C. Dos Santos Ferreira, D.M. Dos Santos Tavares, R.A.P. Rodrigues, Sociodemographic characteristics, functional status and morbidity among older adults with and without cognitive decline, ACTA Paul. Enferm. 24 (1) (2011) 29–35, https://doi.org/10.1590/S0103-21002011000100004.

[18] J. Jang, N. Cushing, L. Clemson, J.R. Hodges, E. Mioshi, Activities of daily living in progressive non-fluent aphasia, logopenic progressive aphasia and Alzheimer's disease, Dement. Geriatr. Cogn. Disord. 33 (5) (2012) 354–360, https://doi.org/10.1159/000339670.

[19] M.E. Mlinac, M.C. Feng, Assessment of activities of daily living, self-care, and Independence, Arch. Clin. Neuropsychol. 31 (6) (2016) 506–516, https://doi.org/10.1093/ARCLIN/ACW049.

[20] A.H. Pinto, C. Lange, C.A. Pastore, P.M.P. de Llano, D.P. Castro, F. dos Santos, Functional capacity to perform activities of daily living among older persons living in rural areas registered in the family health strategy, Cien. Saude Colet. 21 (11) (2016) 3545–3555, https://doi.org/10.1590/1413-812320152111.22182015.

[21] S. Katz, Assessing self-maintenance: activities of daily living, mobility, and instrumental activities of daily living, J. Am. Geriatr. Soc. 31 (12) (1983) 721–727, https://doi.org/10.1111/J.1532-5415.1983.TB03391.X.

[22] D.R. Royall, E.C. Lauterbach, D. Kaufer, P. Malloy, K.L. Coburn, K.J. Black, The cognitive correlates of functional status: a review from the committee on research of the American neuropsychiatric association, J. Neuropsychiatry Clin. Neurosci. 19 (3) (2007) 249–265, https://doi.org/10.1176/JNP.2007.19.3.249.

[23] J.D. Nunes, et al., Indicadores de incapacidade funcional e fatores associados em idosos: estudo de base populacional em Bagé, Rio Grande do Sul, Epidemiol. e Serviços Saúde 26 (2) (2017) 295–304, https://doi.org/10.5123/S1679-49742017000200007.

[24] J. Conti, A interferência dos aspectos percepto-cognitivos nas atividades de vida diária e nas atividades instrumentais de vida diária, em clientes com sequelas por lesão neurológica, Acta Fisiátrica 13 (2) (2006) 83–86, https://doi.org/10.11606/ISSN.2317-0190.V13I2A102588.

[25] D. Kuh, et al., A life course approach to healthy aging, frailty, and capability, J. Gerontol. A Biol. Sci. Med. Sci. 62 (7) (2007) 717–721, https://doi.org/10.1093/GERONA/62.7.717.

[26] Y.W. Zhao, et al., The effect of multimorbidity on functional limitations and depression amongst middle-aged and older population in China: a nationwide longitudinal study, Age Ageing 50 (1) (2021) 190–197, https://doi.org/10.1093/AGEING/AFAA117.

[27] F. Alonso López, Comportamiento adaptativo de las personas con limitación funcional: la adaptación funcional de la vivienda en España, Rev. Esp. Geriatr. Gerontol. 53 (5) (2018) 285–292, https://doi.org/10.1016/J.REGG.2018.02.009.

[28] D.H. Lynch, et al., The relationship between multimorbidity, obesity and functional impairment in older adults, J. Am. Geriatr. Soc. 70 (5) (2022) 1442–1449, https://doi.org/10.1111/JGS.17683.

[29] G. Marques, J. Saini, M. Dutta, P.K. Singh, W.C. Hong, Indoor air quality monitoring systems for enhanced living environments: a review toward sustainable smart cities, Sustain. 12 (10) (2020), https://doi.org/10.3390/SU12104024.

[30] G. Marques, R. Pitarma, N.M. Garcia, N. Pombo, Internet of things architectures, technologies, applications, challenges, and future directions for enhanced living environments and healthcare systems: a review, Electron 8 (10) (2019) 1–27, https://doi.org/10.3390/electronics8101081.

[31] G. Saha, R. Singh, S. Saini, A survey paper on the impact of 'Internet of Things' in healthcare, in: Proc. 3rd Int. Conf. Electron. Commun. Aerosp. Technol. ICECA 2019, 2019, pp. 331–334, https://doi.org/10.1109/ICECA.2019.8822225.

[32] H. Zakaria, N.A. Abu Bakar, N.H. Hassan, S. Yaacob, IoT security risk management model for secured practice in healthcare environment, Procedia Comput. Sci. 161 (2019) 1241–1248, https://doi.org/10.1016/j.procs.2019.11.238.

[33] J.C. Moses, S. Adibi, M. Angelova, S.M.S. Islam, Smart home technology solutions for cardiovascular diseases: a systematic review, Appl. Syst. Innov. 5 (3) (2022) 51, https://doi.org/10.3390/asi5030051.

[34] D. Denyer, D. Tranfield, Producing a systematic review, in: The SAGE Handbook of Organizational Research Methods, 2009, pp. 671–689.

[35] D. Moher, et al., Preferred reporting items for systematic reviews and meta-analyses: the PRISMA statement, PLoS Med. 6 (7) (2009), https://doi.org/10.1371/journal.pmed. 1000097.

[36] M. Ouzzani, H. Hammady, Z. Fedorowicz, A. Elmagarmid, Rayyan-a web and mobile app for systematic reviews, Syst. Rev. 5 (1) (2016) 1–10, https://doi.org/10.1186/s13643-016-0384-4.

[37] No authors listed, Easy as PIE, Nursing 29 (4) (1999) 25.

[38] J.L. Diniz, et al., Internet of things gerontechnology for fall prevention in older adults: an integrative review, Acta Paul. Enferm. 35 (1) (2022) 1–10. [Online]. Available https://search.ebscohost.com/login.aspx?direct=true&db=c8h&AN=155793532&lang=fr&site=ehost-live.

[39] J. Li, W.W. Goh, N.Z. Jhanjhi, F.B.M. Isa, S. Balakrishnan, An empirical study on challenges faced by the elderly in care centres, EAI Endorsed Trans. Pervasive Heal. Technol. 7 (28) (2021) 1–11, https://doi.org/10.4108/eai.11-6-2021.170231.

[40] R. Hu, et al., An unsupervised behavioral modeling and alerting system based on passive sensing for elderly care, Futur. Internet 13 (1) (2021) 1–24, https://doi.org/10.3390/fi13010006.

[41] M. Al-khafajiy, et al., Remote health monitoring of elderly through wearable sensors, Multimed. Tools Appl. 78 (17) (2019) 24681–24706, https://doi.org/10.1007/S11042-018-7134-7/FIGURES/12.

[42] Q. Zhang, et al., The smarter safer Homes solution to support older people living in their own Homes through enhanced care models: protocol for a stratified randomized controlled trial, JMIR Res. Protoc. 11 (1) (2022) 1–23, https://doi.org/10.2196/31970.

[43] L. Correia, et al., Usability of smartbands by the elderly population in the context of ambient assisted living applications, Electron. 10 (14) (2021) 1–18, https://doi.org/10.3390/electronics10141617.

[44] F. Ahamed, S. Shahrestani, H. Cheung, Internet of things and machine learning for healthy ageing: identifying the early signs of dementia, Sensors (Switzerland) 20 (21) (2020) 1–25, https://doi.org/10.3390/s20216031.

[45] F. Tiersen, et al., Smart home sensing and monitoring in households with dementia: user-centered design approach, JMIR Aging 4 (3) (2021), https://doi.org/10.2196/27047.

[46] M.A. Quintana-Suárez, D. Sánchez-Rodríguez, I. Alonso-González, J.B. Alonso-Hernández, A low cost wireless acoustic sensor for ambient assisted living systems, Appl. Sci. 7 (9) (2017) 1–15, https://doi.org/10.3390/app7090877.

[47] F. Nakayama, P. Lenz, M. Nogueira, A resilience management architecture for communication on portable assisted living, IEEE Trans. Netw. Serv. Manag. XX (XX) (2022) 1–13, https://doi.org/10.1109/TNSM.2022.3165729.

[48] D. Loreti, et al., Complex reactive event processing for assisted living: the habitat project case study, Expert Syst. Appl. 126 (2019) 200–217, https://doi.org/10.1016/j.eswa.2019.02.025.

[49] E. Borelli, et al., HABITAT: an IoT solution for independent elderly, Sensors (Switzerland) 19 (5) (2019), https://doi.org/10.3390/s19051258.

[50] A. Grgurić, M. Mošmondor, D. Huljenić, The smarthabits: an intelligent privacy-aware home care assistance system, Sensors (Switzerland) 19 (4) (2019), https://doi.org/10.3390/s19040907.

[51] A. Almeida, R. Mulero, P. Rametta, V. Urošević, M. Andrić, L. Patrono, A critical analysis of an IoT—aware AAL system for elderly monitoring, Futur. Gener. Comput. Syst. 97 (2019) 598–619, https://doi.org/10.1016/j.future.2019.03.019.

[52] L.P. Hung, Y.H. Chao, C.L. Chen, A hybrid key item locating method to assist elderly daily life using internet of things, Mob. Networks Appl. 24 (3) (2019) 786–795, https://doi.org/10.1007/s11036-018-1083-2.

[53] M. Irfan, et al., Non-wearable IoT-based smart ambient behavior observation system, IEEE Sens. J. 21 (18) (2021) 20857–20869, https://doi.org/10.1109/JSEN.2021. 3097392.

[54] G.A. Oguntala, Y.F. Hu, A.A.S. Alabdullah, R.A. Abd-Alhameed, M. Ali, D.K. Luong, Passive RFID module with LSTM recurrent neural network activity classification algorithm for ambient-assisted living, IEEE Internet Things J. 8 (13) (2021) 10953–10962, https://doi.org/10.1109/JIOT.2021.3051247.

[55] W.W.Y. Ng, S. Xu, T. Wang, S. Zhang, C. Nugent, Radial basis function neural network with localized stochastic-sensitive autoencoder for home-based activity recognition, Sensors (Switzerland) 20 (5) (2020), https://doi.org/10.3390/s20051479.

[56] S.M.M. Fattah, I. Chong, Restful web services composition using semantic ontology for elderly living assistance services, J. Inf. Process. Syst. 14 (4) (2018) 1010–1032, https://doi.org/10.3745/JIPS.04.0083.

[57] G. Vallathan, A. John, C. Thirumalai, S.K. Mohan, G. Srivastava, J.C.W. Lin, Suspicious activity detection using deep learning in secure assisted living IoT environments, J. Supercomput. 77 (4) (2021) 3242–3260, https://doi.org/10.1007/s11227-020-03387-8.

[58] A. Caione, A. Fiore, L. Mainetti, L. Manco, R. Vergallo, Top-down delivery of IoT-based applications for seniors behavior change capturing exploiting a model-driven approach, J. Commun. Softw. Syst. 14 (1) (2018) 60–67, https://doi.org/10.24138/jcomss.v14i1.438.

[59] S.A. Moraru, et al., Using IoT assistive technologies for older people non-invasive monitoring and living support in their homes, Int. J. Environ. Res. Public Health 19 (10) (2022), https://doi.org/10.3390/ijerph19105890.

[60] A.V. Román, D.P. Martínez, Á.L. Murciego, D.M. Jiménez-Bravo, J.F. de Paz, Voice assistant application for avoiding sedentarism in elderly people based on iot technologies, Electron. 10 (8) (2021), https://doi.org/10.3390/electronics10080980.

[61] L. Aar, et al., An ambient intelligence-based human behavior monitoring framework for ubiquitous environments, Sensors (Switzerland) 21 (3) (2021) 598–619, https://doi.org/10.3390/s20051479.

[62] N. Thakur, C.Y. Han, An ambient intelligence-based human behavior monitoring framework for ubiquitous environments, Inf. 12 (2) (2021) 81, https://doi.org/10.3390/INFO12020081.

[63] G. Kyriakopoulos, et al., Internet of Things (IoT)-enabled elderly fall verification, exploiting temporal inference models in smart homes, Int. J. Environ. Res. Public Health 17 (2) (2020) 408.

[64] Y.J. Park, S.Y. Jung, T.Y. Son, S.J. Kang, Self-organizing IoT device-based smart diagnosing assistance system for activities of daily living, Sensors (Switzerland) 21 (3) (2021) 1–22, https://doi.org/10.3390/s21030785.

[65] V. Bianchi, M. Bassoli, G. Lombardo, P. Fornacciari, M. Mordonini, I. De Munari, IoT wearable sensor and deep learning: an integrated approach for personalized human activity recognition in a smart home environment, IEEE Internet Things J. 6 (5) (2019) 8553–8562, https://doi.org/10.1109/JIOT.2019.2920283.

[66] A. Bilbao-Jayo, et al., Location based indoor and outdoor lightweight activity recognition system, Electron. 11 (3) (2022) 1–29, https://doi.org/10.3390/electronics11030360.

[67] D. Fuentes, et al., Indoorcare: low-cost elderly activity monitoring system through image processing, Sensors 21 (18) (2021), https://doi.org/10.3390/s21186051.

[68] N. Thakur, C.Y. Han, Multimodal approaches for indoor localization for ambient assisted living in smart homes, Inf. 12 (3) (2021), https://doi.org/10.3390/info12030114.

[69] G. Marques, R. Pitarma, A cost-effective air quality supervision solution for enhanced living environments through the internet of things, Electron. 8 (2) (2019), https://doi.org/10.3390/electronics8020170.

[70] G. Marques, I.M. Pires, N. Miranda, R. Pitarma, Air quality monitoring using assistive robots for ambient assisted living and enhanced living environments through internet of things, Electron. 8 (12) (2019) 1375, https://doi.org/10.3390/electronics8121375.

[71] G. Marques, C.R. Ferreira, R. Pitarma, Indoor air quality assessment using a CO 2 monitoring system based on internet of things, J. Med. Syst. 43 (3) (2019) 67, https://doi.org/10.1007/s10916-019-1184-x.

[72] G. Marques, R. Pitarma, mHealth: indoor environmental quality measuring system for enhanced health and well-being based on Internet of Things, J. Sens. Actuator Netw. 8 (3) (2019) 43, https://doi.org/10.3390/JSAN8030043.

[73] G. Marques, N. Miranda, A.K. Bhoi, B. Garcia-zapirain, S. Hamrioui, I. de la Torre Díez, Internet of Things and enhanced living environments: measuring and mapping air quality using cyber-physical systems and mobile computing technologies, Sensors 20 (3) (2020) 720, https://doi.org/10.3390/S20030720.

[74] G. Marques, R. Pitarma, Particulate matter monitoring and assessment through internet of things: a health information system for enhanced living environments, J. Med. Syst. 44 (12) (2020), https://doi.org/10.1007/s10916-020-01674-8.

[75] E. Gambi, G. Temperini, R. Galassi, L. Senigagliesi, A. De Santis, ADL recognition through machine learning algorithms on IoT air quality sensor dataset, IEEE Sens. J. 20 (22) (2020) 13562–13570, https://doi.org/10.1109/JSEN.2020.3005642.

[76] T. Chaudhuri, Y.C. Soh, H. Li, L. Xie, Machine learning driven personal comfort prediction by wearable sensing of pulse rate and skin temperature, Build. Environ. 170 (2020) 106615, https://doi.org/10.1016/j.buildenv.2019.106615.

[77] A.S. Syed, D. Sierra-Sosa, A. Kumar, A. Elmaghraby, A deep convolutional neural network-XGB for direction and severity aware fall detection and activity recognition, Sensors 22 (7) (2022) 1–22, https://doi.org/10.3390/s22072547.

[78] N. Thakur, C.Y. Han, A study of fall detection in assisted living: identifying and improving the optimal machine learning method, J. Sens. Actuator Netw. 10 (3) (2021) 39, https://doi.org/10.3390/jsan10030039.

[79] N. Gulati, P.D. Kaur, An argumentation enabled decision making approach for fall activity recognition in social IoT based ambient assisted living systems, Futur. Gener. Comput. Syst. 122 (2021) 82–97, https://doi.org/10.1016/j.future.2021.04.005.

[80] M.S. Al-Rakhami, et al., FallDeF5: a fall detection framework using 5G-based deep gated recurrent unit networks, IEEE Access 9 (2021) 94299–94308, https://doi.org/10.1109/ACCESS.2021.3091838.

[81] D. Sarabia-Jácome, R. Usach, C.E. Palau, M. Esteve, Highly-efficient fog-based deep learning AAL fall detection system, Internet of Things (Netherlands) 11 (2020) 100185, https://doi.org/10.1016/j.iot.2020.100185.

[82] A.J. Jara, M.A. Zamora, A.F.G. Skarmeta, An Internet of Things-based personal device for diabetes therapy management in ambient assisted living (AAL), Pers. Ubiquitous Comput. 15 (4) (2011) 431–440, https://doi.org/10.1007/s00779-010-0353-1.

[83] M.K. Hassan, A.I. El Desouky, M.M. Badawy, A.M. Sarhan, M. Elhoseny, M. Gunasekaran, EoT-driven hybrid ambient assisted living framework with naïve Bayes–firefly algorithm, Neural Comput. Applic. 31 (5) (2019) 1275–1300, https://doi.org/10.1007/s00521-018-3533-y.

[84] M.K. Hassan, A.I. El Desouky, S.M. Elghamrawy, A.M. Sarhan, Intelligent hybrid remote patient-monitoring model with cloud-based framework for knowledge discovery, Comput. Electr. Eng. 70 (2018) 1034–1048, https://doi.org/10.1016/j.compeleceng.2018.02.032.

[85] M. Tariq, H. Majeed, M.O. Beg, F.A. Khan, A. Derhab, Accurate detection of sitting posture activities in a secure IoT based assisted living environment, Futur. Gener. Comput. Syst. 92 (2019) 745–757, https://doi.org/10.1016/j.future.2018.02.013.

[86] B. Andò, et al., An introduction to patterns for the internet of robotic things in the ambient assisted living scenario, Robotics 10 (2) (2021) 1–18, https://doi.org/10.3390/robotics10020056.

[87] S. Tanabe, et al., Designing a robotic smart home for everyone, especially the elderly and people with disabilities, Fujita Med. J. 5 (2) (2019) 31–35, https://doi.org/10.20407/fmj.2018-009.

[88] L. Lam, L. Fadrique, G. Bin Noon, A. Shah, P.P. Morita, Evaluating challenges and adoption factors for active assisted living smart environments, Front. Digit. Heal. 4 (May) (2022) 1–10, https://doi.org/10.3389/fdgth.2022.891634.

[89] D. Facchinetti, G. Psaila, P. Scandurra, Mobile cloud computing for indoor emergency response: the IPSOS assistant case study, J. Reliab. Intell. Environ. 5 (3) (2019) 173–191, https://doi.org/10.1007/s40860-019-00088-9.

[90] D. Choi, H. Choi, D. Shon, Future changes to smart home based on AAL healthcare service, J. Asian Archit. Build. Eng. 18 (3) (2019) 194–203, https://doi.org/10.1080/13467581.2019.1617718.

[91] K. Witrisal, et al., High-accuracy localization for assisted living: 5G systems will turn multipath channels from foe to friend, IEEE Signal Process. Mag. 33 (2) (2016) 59–70, https://doi.org/10.1109/MSP.2015.2504328.

[92] L. Kanaris, A. Kokkinis, A. Liotta, S. Stavrou, Fusing bluetooth beacon data with Wi-Fi radiomaps for improved indoor localization, Sensors (Switzerland) 17 (4) (2017) 1–15, https://doi.org/10.3390/s17040812.

[93] M. Kolakowski, Improving accuracy and reliability of Bluetooth low-energy-based localization systems using proximity sensors, Appl. Sci. 9 (19) (2019), https://doi.org/10.3390/app9194081.

[94] F. Addante, F. Gaetani, L. Patrono, D. Sancarlo, I. Sergi, G. Vergari, An innovative AAL system based on IoT technologies for patients with sarcopenia, Sensors (Switzerland) 19 (22) (2019) 1–18, https://doi.org/10.3390/s19224951.

[95] A.J. Jara, M.A. Zamora, A.F. Skarmeta, Drug identification and interaction checker based on IoT to minimize adverse drug reactions and improve drug compliance, Pers. Ubiquitous Comput. 18 (1) (2014) 5–17, https://doi.org/10.1007/s00779-012-0622-2.

[96] M.I. Ahmed, G. Kannan, Safeguards and weightless of electronic chain of command consolidated for virtual patient evaluation, Multimed. Tools Appl. (2022), https://doi.org/10.1007/s11042-022-13310-3.

[97] L.P. Hung, W. Huang, J.Y. Shih, C.L. Liu, A novel IoT based positioning and shadowing system for dementia training, Int. J. Environ. Res. Public Health 18 (4) (2021) 1–21, https://doi.org/10.3390/ijerph18041610.

[98] A.S. Morales, F. de Oliveira Ourique, L.D. Morás, S.C. Cazella, Exploring interpretable machine learning methods and biomarkers to classifying occupational stress of the health workers, Intell. Syst. Ref. Libr. 121 (2022) 105–124, https://doi.org/10.1007/978-3-030-97516-6_6/COVER/.

About the authors

Analúcia Schiaffino Morales is PhD in Electrical Engineering from the Federal University of Santa Catarina, UFSC (2003). Professor in the Computer Department at UFSC and postgraduate program in Energy and Sustainability. Member of the Research Group on Intelligent Systems Applied to Health. She has experience in national and international project partnerships in smart systems, Web 3.0 and IoT applications. Areas of interest are: IoT applied to Health and sustainability, decision support systems, explainable AI, and intelligent systems.

Ione Jayce Ceola Schneider is a Physiotherapist and PhD in Public Health. She is a professor in the Physiotherapy Undergraduate Course and in the Postgraduate Program in Rehabilitation Sciences and Public Health at the Federal University of Santa Catarina.

Fabrício de Oliveira Ourique received the B.Sc. degree in 2000 from the Pontifical Catholic University of Rio Grande do Sul, Porto Alegre, RS, Brazil, the MSc degree in 2002 from the University of New Mexico, Albuquerque, NM, United States, and the PhD degree in 2005 from the University of New Mexico, Albuquerque, NM, United States, all in Electrical Engineering. In 2006, he was Assistant Professor at Catholic University of Rio Grande do Sul. In 2012, he joined Federal University of Santa Catarina, SC, Brazil, where he is currently an Associate Professor of Computer Department. His research interests lie in the area of signal and image processing and statistical signal processing.

Silvio César Cazella is PhD in Computer Science at the Federal University of Rio Grande do Sul, having completed one year of this PhD at the University of Alberta in Canada. Master in Computer Science from the Federal University of Rio Grande do Sul. He is currently Associate Professor-Level III at the Federal University of Health Sciences in Porto Alegre/Brazil. Permanent researcher in the graduate Programs in Information Technologies and Health Management and Health Education, and Collaborate Researcher in the graduate Program in Health Sciences. Member of the Innovation Center on Artificial Intelligence for Health (CIIA-Health) and Center for Artificial Intelligence Applied to Health (CIARS) in Brazil. Topics of Interest: Artificial Intelligence, Recommender Systems, Multi-Agent Systems, Internet of Things and Internet of Medical Things. Also noteworthy is Data Mining in the field of Health and Education as well as Business Intelligence and Machine Learning.

CHAPTER SEVEN

Internet of things and data science methods for enhanced data processing

Pan Zheng[a] and Bee Theng Lau[b]
[a]Department of Accounting and Information Systems, University of Canterbury, Christchurch, New Zealand
[b]Faculty of Engineering, Computing and Science, Swinburne University of Technology Sarawak, Sarawak, Malaysia

Contents

Abstract

The Internet of Things (IoT) is a rapidly expanding field of research and application that focuses on creating networks of physical objects embedded with sensors, software, and connectivity. The number of IoT devices continues to grow, there is a corresponding surge in the amount of data they generate. This data presents opportunities for gaining valuable insights and making better decisions across various sectors, such as healthcare, agriculture, business, logistics, and manufacturing. However, the growing volume, speed, and diversity of IoT data present significant challenges for processing and analysis using traditional methods. This is where the role of data science methods becomes crucial. In this chapter, we examine existing frameworks and overarching methodologies widely employed in the field of data science and data analytics. Drawing inspiration from these, we introduce the Data Science Life Cycle for IoT Applications (DSLC-IoT), which serves as a fundamental guideline for addressing data-intensive and data-driven challenges in the realm of IoT. Subsequently, we delve into an exploration of various data science methods utilized to enhance data processing within the context of IoT. The review centers on the application areas of healthcare, agriculture, business, logistics, and manufacturing. These areas have significant social and economic importance and are closely intertwined with people's lives.

1. Introduction

The Internet of Things (IoT) is a rapidly growing research and application field that aims to develop networks of physical objects embedded with sensors, software, and connectivity, which enables them to exchange data with other devices and systems over the Internet. With the proliferation of IoT devices, there is an exponential increase in the amount of data generated by these devices. This data can be harnessed to gain insights and improve decision-making in various domains, including healthcare, agriculture, business, logistics and manufacturing. The sheer volume, velocity, and variety of IoT data pose significant challenges for data processing and analysis. Traditional data processing techniques are inadequate to handle the complexity and scale of IoT data. This is where data science methods come into play. Data science involves the application of statistical, mathematical, and machine learning techniques to extract insights from large and complex datasets. In this context, data science methods have become essential for processing and analyzing IoT data. By leveraging data science methods, it is possible to extract meaningful insights from IoT data, predict future outcomes, and optimize processes. This can lead to improved decision-making, enhanced efficiency, cost savings and better computational results.

In this chapter, we first go through some prevalent and renowned frameworks and high-level approaches for data science and data analytics. Enlightened by those, we proposed Data Science Life Cycle for IoT Applications (DSLC-IoT), which can serve as a based line for data-intensive and data-driven IoT problems. Subsequently, we explored some of the data science methods used for enhanced data processing in IoT. We reviewed data science methods that can be applied to IoT data to gain insights and improve decision-making in three primary application areas of the IoT that hold immense social and economic value: Healthcare IoT, Agriculture IoT, and Business IoT. These areas are investigated in-depth, covering a broad range of real-life issues. Each of these domains has notable and representative research studies that shed light on the data science methods currently in use.

2. Data science and analytics life cycle

This section provides a description of the major data analytics frameworks and approaches. These approaches may offer high-level guidance when IoT data is processed and analyzed. Previously, researchers proposed different approaches and frameworks of data analytics [1–5] for problems in different domains. It is a concept that has changed over time and has been affected by numerous disciplines and practices, hence there is no standardized data analytics life cycle that suits all situations.

2.1 Cross-industry standard process for data mining

CRISP-DM (Cross-Industry Standard Process for Data Mining) [6] is one of the most popular frameworks for planning and executing data science and data mining projects. It is an open standard that offers an organized and methodical approach to the entire data mining process, from problem understanding and data preparation to the evaluation and deployment of models. It was proposed by the CRISP-DM consortium which consisted of three major players in the data processing and data science industry, DaimlerChrysler AG, SPSS Inc. and NCR Corporation (Fig. 1).

The first phase is Business Understanding. Data analytics always starts with a practical problem. The "business" here can be generally interpreted as the environment of the problem. In fact, it can be in the field of engineering and science. This phase involves understanding the actual problem including its scope, background and constraints, and more importantly, its objectives. As the first phase in the life cycle, a data mining project/task plan should be developed to better illustrate the needs of different stakeholders, data and resource requirements and the expected outcomes at a high level.

In the Data understanding phase, the relevant data is identified, collected, and explored to understand its quality, completeness, and relevance to the data mining problem. There are often four tasks in this phase, namely data collection, data description, data assessment and data exploration. Data collection means that data are obtained and gathered from various sources based on data mining problems. Data description is related to the tasks that investigate the structures, types, relationships and distributions of the data. It also identifies the values that are most pertinent to the data mining problem. Initial data exploration can be carried out to better

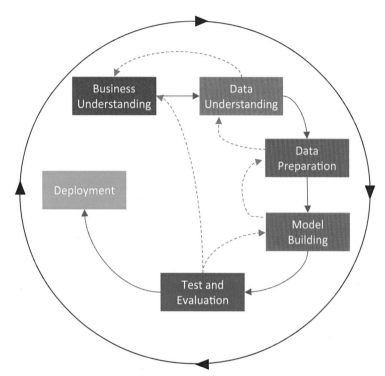

Fig. 1 The six-phase CRISP-DM data mining life cycle.

understand the data, especially to find the trends, patterns, outliers and anomalies of the data. Data assessment is about examining and verifying the data, which includes evaluating the quality, completeness, and relevance of the data.

Data preparation is the phase focusing on data pre-processing, data cleansing and data wrangling [7]. Data cleaning task is carried out by removing or imputing missing values, correcting inconsistent or incorrect values, and removing duplicates or irrelevant records. Data pre-processing involves data consolidation, feature extraction, data balancing and dimension reduction. Data wrangling is a process of tailoring and mapping raw data which can be better used by the consumer application. By completing these tasks, the analyst prepares the data for analysis and ensures that it is of high quality and suitable for the chosen modeling techniques. This, in turn, helps to improve the accuracy and reliability of the results obtained from the analysis.

Model building involves selecting and applying appropriate models or algorithms to the prepared data. Depending on the type of analytics, modeling techniques, such as linear regression, decision trees, neural networks, optimization or simulation, can be selected. Model training, model evaluation and validation, and model refinement are the major activities that happen in this phase.

Model testing and evaluations are closely related to the model-building phase. In this phase, the model's effectiveness and efficiency are evaluated based on its ability to meet the business and data mining goals defined in the first phase. This involves testing the model on new data sets and comparing its performance with other models. Metrics, e.g., Mean Squared Error, R-squared, and Area Under the ROC Curve, are used to determine how well the model is able to predict and classify.

After the testing and evaluation, it is the last phase of CRISP-DM, where the model is integrated into the data mining solution. The model's ongoing performance is monitored to ensure that it continues to meet the requirements. The monitoring metrics include accuracy, throughput, latency and robustness. If issues or anomalies are identified, timely adjustments and updates are implemented. CRISP-DM is a comprehensive and iterative process model that covers the entire data mining process, from problem understanding and data understanding to evaluation and deployment. It provides a structured approach for data mining projects, ensuring that all relevant aspects of the project are taken into account. It also exhibits a good deal of adaptability. It is flexible enough to be applied to several kinds of data mining initiatives, regardless of the sector or target audience. It offers a broad framework that can be altered to meet the particular requirements of the project. For data mining, it is a widely recognized and approved industry standard. Many businesses, including finance, healthcare, retail, and telecommunications, have effectively used it. Moreover, it also demonstrates a good deal of flexibility. it can be adapted to various types of data mining projects and problems, regardless of the industry or application area. It offers a general framework that can be customized to fit the specific needs of the project. it is a widely accepted and recognized industry standard for data mining. It has been used successfully in various industries.

2.2 Knowledge discovery in databases

KDD (Knowledge Discovery in Databases) [8] was a series of processes that were initially designed to generate knowledge and discover patterns in structured data stored in databases, especially relational databases. KDD involves a sequence of steps that includes data selection, cleaning, integration, transformation, mining, pattern evaluation, and knowledge representation. These steps are often referred to as the KDD processes. With the advancement of data science, many approaches to data science problems take inspiration from KDD processes. It is worthwhile to look at KDD here.

There are five processes in KDD, which are selection, pre-processing, transformation, data mining and interpretation/evaluation. In the selection process, the relevant data is selected from a larger set of data sources based on the requirement of the data science problem. The pre-processing is very similar to data preparation in CRISP-DM, in which the selected data is cleaned and integrated. In the data transformation step, relevant data is transformed into a suitable format for further analysis. Data mining is the process of drawing patterns and knowledge from data using statistical and machine learning methods. Domain specialists then interpret the patterns that were identified in order to gather knowledge and make informed decisions. The last process of KDD is Interpretation that the gain insights and results are explained and communicated with relevant parties.

2.3 Sample, explore, modify, model, and assess (SEMMA)

SEMMA [9] is a data mining methodology developed by SAS Institute that provides a framework for data mining projects. The five phases of the approach are referred to by the acronym SEMMA, which are Sample, Explore, Modify, Model and Assess. In "Sample" phase, a subset of the data is obtained for the large dataset. Often the selected data should bear distinctive features and characters of the whole dataset to ensure the soundness of the analytical results. The second phase is "Explore" which is quite similar to the data understanding of the CRISP-DM. It finds and explores the distribution of the data values, relationships among different variables, outliers and predominant features. The third phase is "Modify" which can be mapped to the data preparation of CRISP-DM. It includes pre-analytics activities such as data cleansing and data wrangling. In the "Model" phase, statistical, machine-learning models are usually developed based on the data and the problem of study. it is similar to the model building of CRISP-DM. The final phase is "Assess." The performance of the model is evaluated to determine its accuracy and effectiveness. This involves testing the model on a validation set and using metrics such as accuracy to evaluate its performance.

2.4 Data science life cycle for IoT applications

The IoT is a fast-expanding field with the potential to transform a variety of aspects of the way how we live. IoT applications often generate a vast amount of data and it is crucial to have a dedicated data science and analytics life cycle for IoT-related study and research. With reference to various data science and mining approaches, hereby the data science life cycle for IoT applications (DSLC-IoT) is proposed (Fig. 2).

IoT Application Problem is the first phase, where the problem of interest is identified and formulated. Relevant background, constraints and requirements are identified in this phase. The problem in IoT applications may involve a wide range of use cases, from optimizing energy usage in a building to predicting equipment failure in a manufacturing plant. The second phase of the life cycle is IoT Data understanding and collection. In this stage, data is collected from various sources such as sensors, devices, and other sources of IoT data. The data is then consolidated and integrated into a single data store, e.g., data lake. In this phase, some preliminary study is carried out to interpret and make sense of the data, e.g., the structure of data, the dimensions of the data and the categories of the data. In the data preparation and wrangling phase, the collected data is cleaned, transformed, and formatted

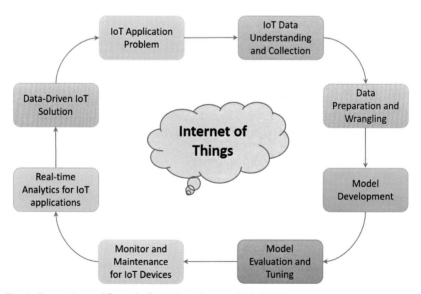

Fig. 2 Data science life cycle for IoT application (DSLC-IoT).

into a consistent and usable format. It involves removing duplicates, filling in missing data, and converting data into a format that can be analyzed. Model development is related to the use of statistical, machine learning and AI methods and techniques to create data science models. Depending on the IoT problem, the models can be used for making predictions or classifications The IoT devices are monitored for performance, data quality, and reliability in the "Monitor & Maintenance for IoT Devices" phase. Maintenance and repair of devices are carried out as needed. it is important to take the IoT hardware aspects in the life cycle. IoT hardware is the essential component to enable the functionality of IoT systems. For example, IoT hardware such as sensors and monitoring systems can be used to collect data about the performance of machines and equipment. This data is then analyzed using data science models such as predictive maintenance to predict when maintenance is required, reducing downtime and improving operational efficiency. IoT devices often play the data source role. In most IoT systems, the capability of analyzing data in real-time is needed. Stream processing is used to process and analyze high-velocity data streams. IoT hardware such as edge computing devices can be used to process data in real time at the edge of the network. This allows for faster processing of data and reduces the amount of data that needs to be transmitted to the cloud for analysis, reducing latency and improving the overall performance of the IoT system. The last phase is data-driven IoT solution. Eventually, the data science models need to serve the purpose of IoT problem-solving and sometimes decision-making. By collecting and analyzing data from the environment, businesses can make data-driven decisions, improve operational efficiency, and provide more accurate results to IoT application problems.

3. Data science in healthcare IoT

Data science methods and techniques have significant applications in healthcare IoT (Internet of Things) problems [10–12]. Patient Monitoring [13,14] using Healthcare IoT devices such as wearables, sensors, and mobile health apps collect real-time data on patient health metrics such as heart rate, temperature, blood pressure, and oxygen saturation. Data science algorithms can analyze this data to identify trends and predict health risks, enabling clinicians to provide more personalized care. Such application can be further extended in a remote scenario [15].

Machine learning is a powerful tool for analyzing data from healthcare IoT devices and can be used to develop more personalized and effective treatments, identify new patient subgroups, and improve patient outcomes. In healthcare IoT, supervised learning can be used to predict patient outcomes or classify patient conditions based on input data from IoT devices such as wearables, sensors, or medical imaging. Reinforcement learning can be used to optimize treatment plans for patients with chronic conditions, such as diabetes or hypertension. Unsupervised learning can be used to identify new patient subgroups or to cluster patients with similar conditions based on data from IoT devices. Machine learning algorithms can be used to analyze large amounts of data from healthcare IoT devices, such as wearables or sensors, to identify patterns and make predictions about patient health. For example, machine learning algorithms can be used to analyze data from glucose monitors to predict hypoglycemic events in patients with diabetes [16].

Data mining is another key method of data science used in IoT applications and systems. Association rule mining involves finding patterns in data that occur together frequently. In healthcare IoT, association rule mining can be used to identify relationships between patient symptoms and conditions, or between patient characteristics and treatment outcomes [17,18]. Clustering involves grouping similar data points together based on their attributes. In healthcare IoT, clustering can be used to identify patient subgroups with similar characteristics, such as risk factors for a particular condition or response to treatment [19,20]. Classification is about assigning data points to predefined categories based on their attributes. In healthcare IoT, classification can be used to predict patient outcomes or to identify patients with a particular condition based on data from IoT devices [21]. Regression involves analyzing the relationship between variables to predict a continuous outcome. In healthcare IoT, regression methods can be used to predict patient outcomes, such as the risk of readmission to the hospital or the progression of a particular condition [22]. Outlier detection involves identifying data points that are significantly different from the rest of the dataset. In healthcare IoT, outlier detection can be used to identify patients who are at high risk of a particular condition or who may require more intensive care [23]. Decision trees involve creating a tree-like structure that represents decisions and their possible consequences. In healthcare IoT, decision trees can be used to predict patient outcomes or to guide treatment decisions based on data from IoT devices [24]. Principal component analysis transforms high-dimensional data into a lower-dimensional space while retaining as much information as possible. In healthcare IoT, principal component analysis can be used to identify key features or variables that are most relevant for predicting patient outcomes or classifying patient conditions [25]. Text mining extracts valuable insights from unstructured text data, such as patient notes or medical records. In healthcare IoT, text mining can be used to identify patterns in patient data that may be missed by structured data analysis.

In healthcare IoT, data science can be used to analyze patient data and predict the onset of diseases, identify risk factors, and personalize treatment plans. It can also be used to optimize hospital operations, improve patient safety, and reduce readmission rates. Moreover, data science can play a crucial role in the development of new medical devices and drugs by enabling efficient clinical trials and facilitating drug discovery. However, there are still several challenges to be addressed [26], such as data security and privacy concerns, data quality issues, and the need for collaboration between healthcare providers, data scientists, and IoT engineers. Nevertheless, with the continued advancement of technology and the increasing adoption of IoT devices in healthcare, the future of data science in healthcare IoT looks promising. How data science methods are used in the healthcare sector could be revolutionary because of IoT. Data science techniques such as machine learning, data mining, and predictive analytics can be used to improve diagnosis outcomes, lower costs, and boost the effectiveness of healthcare services as a result of the explosive rise of IoT devices and the growing availability of healthcare data.

4. Data science in agriculture IoT

The use of IoT devices and sensors in agriculture to maximize crop yields, decrease waste, and enhance efficiency is known as agriculture IoT, also known as Smart Agriculture. Soil sensors, weather stations, GPS trackers, drones, and smart irrigation systems are a few IoT instruments utilized in agriculture. By collecting and analyzing data collected by various IoT networks and devices, e.g. soil moisture, temperature, weather patterns, and other environmental factors, farmers can make more informed decisions about planting, fertilization, and irrigation, leading to increased crop yields and reduced impacts of natural disasters. Data science and analytics methods play important roles in those studies and processes.

Predictive analytics and modeling use statistical methods and machine learning algorithms to predict future outcomes. In Agriculture IoT, predictive analytics can be used to forecast crop yields, predict weather patterns, and identify potential crop diseases before they occur [27,28]. Regression is one of the popular predictive analytics methods often used. For example, Elastic Net Regression and Multiple Linear Regression are used to predict soil moisture based on the data collected from IoT devices [29]. Machine learning based on analytics approaches is used to carry out those studies, e.g., a fuzzy logic-based model is constructed to analyze the weather condition [30]. A study of real-time analysis of weather parameters is presented [31–33]. For crop yields, a multilayer perceptron neural network (MLP), a machine learning method, is used to predict sugarcane yield production in an IoT environment [34].

The time factor contributes significantly to Agriculture IoT, hence in agriculture IoT, time-series analysis can be utilized to identify trends and patterns in temperature [35], humidity [36], rainfall [37], soil conditions [38] and wind speed [37] captured by IoT sensors. Most of the work focus on descriptive and predictive analytics.

Classification and data-driven agriculture decision-making is another important problem in agriculture IoT. Decision tree methods exhibit the capability in those problems. The decision tree can be used for data fusion, where a comprehensive view is created based on the data from multiple IoT sensors [39]. Classifying farming machinery is another application of the decision tree method [40]. The paradigms are developed for classifying different environmental factors of agriculture IoT, Here are some studies of such problems. A Gradient Boost Decision Tree is created to model the internal temperature of a greenhouse [41]. A Decision tree is developed to predict the health situation of crops based on environmental factors [42,43]. For the honey and bee products industry, to monitor the health condition of the bees, a decision tree model is created to classify bee hive conditions [44]. Similarly, we found the use of a decision tree in monitoring and classifying cow behaviors [45]. Decision Tree also finds its applications in crop yield maximization and prediction [46],

The agriculture industry has undergone a significant transformation with the integration of data science through the use of IoT sensors. This technology allows farmers to collect vast amounts of data related to crop growth, soil conditions, and weather patterns. By applying advanced data analytics and machine learning techniques, farmers can gain valuable insights into their farming practices. Leveraging IoT-enabled sensors, farmers can collect massive amounts of data about their crops, soil, weather conditions, livestock health and more, and use data science and related methods to gain valuable insights for their farming operations. As the world's population continues to grow and the demand for food increases, the importance of data science in agriculture IoT is set to rise. It is a powerful instrument that enables farmers to feed the world while minimizing negative impacts on the environment. The future of agriculture, including animal husbandry and aquaculture, depends on better understanding of the farming environment and relevant factors, where data science methods facilitate such understanding. Data science methods will continue to be a driving force in shaping the industry for years to come.

5. Data science in business IoT

IoT can offer numerous benefits and improve business processes. IoT devices in the context of business can generate vast amounts of data from various sources such as customer interactions, manufacturing processes, and supply chain operations. Data science methods and techniques can be used to extract meaningful information from this data and to develop computational models that can help businesses make data-driven decisions. There have been loads of studies on IoT models for business, business management and business operation proposed by different researchers [47–50]. This section looks at the data sciences methods for solving specific business-related problems. IoT explores its potential uses in inventory management, Supply Chain Management (SCM) and optimization, predictive maintenance of equipment and facilities, workplace safety monitoring, customer experience and relationship management(CRM). Among all these practical problems, we choose a few important aspects and look at the data science methods used in providing solutions to the problems.

IoT is being used more and more in inventory management to increase warehouse efficiency, asset tracking and logistics operations. Data science methods Naive Bayes, K-nearest neighbors and Support Vector Machine are used in SKU (stock keeping unit) classification of items for better inventory management [51]. Manufacture demand forecast is an important problem for inventory management. Multiple linear regressions (MLR) with genetic algorithms (GA) is used to make such predictions [52]. Machine learning algorithms are used to understand the operation status of the power tools and assets with IoT SmartTag [53]. Indoor positioning using IoT devices can greatly help inventory operations. Multi-layer perceptron [54] and linear regression method [55] are two methods employed in indoor positioning studies. Supply Chain Management and optimization is an application domain closely related to inventory management. Harnessing the potential of IoT technologies, supply chain managers can gain real-time visibility into their operations and make data-driven decisions that drive efficiency and improve profitability. The statistical approach Peak Over Threshold (POT) method of the extreme value theory is used to supply chain risk based on IoT [56]. particle swarm optimization (PSO) algorithm is used to track, trace and manage the operations of food supply chain management.

There are some studies carried out on predictive maintenance using IoT for business. IoT sensors can be attached to machinery and capture the status of machines. The data obtained can be used for predicting maintenance needs [57]. Exploratory data analysis using Classification and Regression Trees(CART), Neural Networks and Support Vector Machines to train a classifier to detect quality failures in production cycles. Autoregressive Integrated Moving Average (ARIMA) predict the future status of machines. An Improved Consensus self-organized model is found in IoT-based predictive maintenance [58]. The Learning Remaining-Useful-Time (LRUT) is used in predictive maintenance based on the equipment vibration data captured by IoT sensors [59].

Customer experience and customer relationship management (CRM) have become key differentiators in today's competitive market [60]. IoT has enabled businesses to personalize their customer experiences. With the ability to track customer behavior and preferences, businesses can tailor their products and services to better meet their customers' needs which leads to more personalized interactions and improved customer satisfaction. An open and automated customer service platform based on IoT, blockchain, and automated machine learning is proposed [61]. AutoML, one of the major components of the technical layer of the framework, is implemented to process a large amount of customer service data collected using IoT sensors. Customer behavior classification using three different clustering methods [62], agglomerative hierarchical clustering, K-Medoids clustering and Density-Based Spatial Clustering of Applications with Noise (DBSCAN), where the first two methods seem to perform better. Classification and Regression Trees (CART) is used to predict sales of different categories of items depending on the customer activities at convenience stores in gas stations [63].

Data Science methods are rapidly transforming the way businesses leverage IoT technologies to drive growth and innovation. The power of IoT lies in the ability to collect and analyze vast amounts of data, but it is

Data Science that provides the tools and methods to extract actionable insights from this data. By adopting a Data Science-driven approach to IoT, businesses can gain a competitive advantage by improving their operations, creating new revenue streams, and delivering personalized experiences to their customers. The successful integration of Data Science and IoT will require a strategic approach that considers the unique needs and goals of each business, as well as the rapidly evolving landscape of technology and data.

6. Conclusion

Data science methods are essential for making sense of the vast amounts of data generated by IoT devices. Advanced data science algorithms and machine learning models can identify patterns, anomalies, and correlations in the data, enabling businesses to make data-driven decisions. In this chapter, different data science life cycles and frameworks for general analytical problems are reviewed first to reflect the state-of-the-art. The current prevalent frameworks are mostly iterative multiphase approaches. Based on those structured approaches, we proposed a Data Science Life Cycle for IoT Applications (DSLC-IoT), which aims to provide a high-level architecture for practical data-driven IoT problems. Three major IoT application areas with high social and economic values, namely Healthcare IoT, Agriculture IoT and Business IoT, are closely investigated with a wide spectrum of real-life problems. In each of the areas, we find some notable and representative studies which enable us to explore the data science methods in use. It is found that statistical methods and conventional AI methods are used in earlier literature, whereas in the latest data-driven IoT problems, machine learning and neural network methods are increasingly gaining popularity, which, in fact, reflect the current trend of machine/deep learning development and application in all science and engineering fields. However, the use of IoT and data science methods raises concerns regarding data privacy and security. The vast amounts of data collected by IoT devices can include sensitive information such as personal health information, financial information, and location data. As such, organizations need to take steps to ensure that this data is secure and protected from unauthorized access or breaches.

The integration of IoT and data science offers exciting opportunities for enhanced data processing. As the technology continues to advance, businesses and industries need to stay abreast of the latest developments to maximize the benefits of IoT and data science while minimizing the risks. With proper planning and execution, the combination of IoT and data science can transform industries, improve operations, and enhance the quality of life for individuals around the world.

References

[1] D. Dietrich, B. Heller, B. Yang, Data Science & Big Data Analytics: Discovering, Analyzing, Visualizing and Presenting Data, Wiley, 2015.

[2] N.W. Grady, J.A. Payne, H. Parker, Agile big data analytics: AnalyticsOps for data science, in: 2017 IEEE International Conference on Big Data (Big Data), IEEE, 2017, pp. 2331–2339.

[3] M. Schulz, U. Neuhaus, J. Kaufmann, D. Badura, S. Kuehnel, W. Badwitz, D. Dann, S. Kloker, E.M. Alekozai, C. Lanquillon, Introducing DASC-PM: A Data Science Process Model, ACIS 2020 Proc. 45 (2020). https://aisel.aisnet.org/acis2020/45.

[4] K. Rahul, R.K. Banyal, Data life cycle management in big data analytics, Procedia Comput. Sci. 173 (2020) 364–371.

[5] H.-N. Dai, R.C.-W. Wong, H. Wang, Z. Zheng, A.V. Vasilakos, Big data analytics for large-scale wireless networks: challenges and opportunities, ACM Comput. Surv. 52 (5) (2019) 1–36.

[6] P. Chapman, J. Clinton, R. Kerber, T. Khabaza, T. Reinartz, C. Shearer, R. Wirth, CRISP-DM 1.0: Step-by-step data mining guide, SPSS Inc. 9 (13) (2000) 1–73.

[7] F. Endel, H. Piringer, Data Wrangling: Making data useful again, IFAC-PapersOnLine 48 (1) (2015) 111–112.

[8] U.M. Fayyad, G. Piatetsky-Shapiro, P. Smyth, Knowledge discovery and data mining: towards a unifying framework, in: E. Simoudis, J. Han, U. Fayyad (Eds.), Proceedings of the Second International Conference on Knowledge Discovery and Data Mining, AAAI Press, 1996, pp. 82–88.

[9] R. Matignon, Data Mining Using SAS Enterprise Miner, John Wiley & Sons, 2007.

[10] S.B. Baker, W. Xiang, I. Atkinson, Internet of things for smart healthcare: technologies, challenges, and opportunities, IEEE Access 5 (2017) 26521–26544.

[11] S. Selvaraj, S. Sundaravaradhan, Challenges and opportunities in IoT healthcare systems: a systematic review, SN Appl. Sci. 2 (1) (2020) 139.

[12] W. Li, Y. Chai, F. Khan, S.R.U. Jan, S. Verma, V.G. Menon, X. Li, A comprehensive survey on machine learning-based big data analytics for IoT-enabled smart healthcare system, Mob. Netw. Appl. 26 (2021) 234–252.

[13] M.A. Akkaş, R. Sokullu, H.E. Çetin, Healthcare and patient monitoring using IoT, Internet of Things 11 (2020) 100173.

[14] H. Pandey, S. Prabha, Smart health monitoring system using IOT and machine learning techniques, in: 2020 Sixth International Conference on Biosignals, Images, and Instrumentation (ICBSII), IEEE, 2020, pp. 1–4.

[15] H.T. Yew, M.F. Ng, S.Z. Ping, S.K. Chung, A. Chekima, J.A. Dargham, IOT based real-time remote patient monitoring system, in: 2020 16th IEEE International Colloquium on Signal Processing & its Applications (CSPA), IEEE, 2020, pp. 176–179.

[16] G. Alfian, M. Syafrudin, N.L. Fitriyani, M.A. Syaekhoni, J. Rhee, Utilizing IoT-based sensors and prediction model for health-care monitoring system, in: Artificial Intelligence and Big Data Analytics for Smart Healthcare, Academic Press, 2021, pp. 63–80.

[17] J. Nahar, T. Imam, K.S. Tickle, Y.P.P. Chen, Association rule mining to detect factors which contribute to heart disease in males and females, Expert Syst. Appl. 40 (4) (2013) 1086–1093.

[18] M. Tandan, Y. Acharya, S. Pokharel, M. Timilsina, Discovering symptom patterns of COVID-19 patients using association rule mining, Comput. Biol. Med. 131 (2021) 104249.

[19] A. Srivastava, A. Singh, S.G. Joseph, M. Rajkumar, Y.D. Borole, H.K. Singh, WSN-IoT clustering for secure data transmission in E-health sector using green computing strategy, in: 2021 9th International Conference on Cyber and IT Service Management (CITSM), IEEE, 2021, pp. 1–8.

[20] J.A. Onesimu, J. Karthikeyan, Y. Sei, An efficient clustering-based anonymization scheme for privacy-preserving data collection in IoT based healthcare services, Peer Peer Netw. Appl. 14 (2021) 1629–1649.

[21] A.U. Haq, J.P. Li, S. Khan, M.A. Alshara, R.M. Alotaibi, C. Mawuli, DACBT: deep learning approach for classification of brain tumors using MRI data in IoT healthcare environment, Sci. Rep. 12 (1) (2022) 15331.

[22] D. Upadhyay, P. Garg, S.M. Aldossary, J. Shafi, S. Kumar, A linear quadratic regression-based synchronised health monitoring system (SHMS) for IoT applications, Electronics 12 (2) (2023) 309.

[23] M.A. Samara, I. Bennis, A. Abouaissa, P. Lorenz, A survey of outlier detection techniques in IoT: review and classification, J. Sens. Actuator Netw. 11 (1) (2022) 4.

[24] R. Manikandan, R. Patan, A.H. Gandomi, P. Sivanesan, H. Kalyanaraman, Hash polynomial two factor decision tree using IoT for smart health care scheduling, Expert Syst. Appl. 141 (2020) 112924.

[25] J.M. Philip, S. Durga, D. Esther, Deep learning application in iot health care: a survey, in: Intelligence in Big Data Technologies—Beyond the Hype: Proceedings of ICBDCC 2019, Springer Singapore, 2021, pp. 199–208.

[26] R. De Michele, M. Furini, IoT healthcare: benefits, issues and challenges, in: Proceedings of the 5th EAI International Conference on Smart Objects and Technologies for Social Good, 2019, pp. 160–164.

[27] M. Lee, J. Hwang, H. Yoe, Agricultural production system based on IoT, in: 2013 IEEE 16Th International Conference on Computational Science and Engineering, IEEE, 2013, pp. 833–837.

[28] L.O. Colombo-Mendoza, M.A. Paredes-Valverde, M.D.P. Salas-Zárate, R. Valencia-García, Internet of things-driven data mining for smart crop production prediction in the peasant farming domain, Appl. Sci. 12 (4) (2022) 1940.

[29] G. Singh, D. Sharma, A. Goap, S. Sehgal, A.K. Shukla, S. Kumar, Machine learning based soil moisture prediction for internet of things based smart irrigation system, in: 2019 5th International Conference on Signal Processing, Computing and Control (ISPCC), IEEE, 2019, pp. 175–180.

[30] B. Keswani, A.G. Mohapatra, A. Mohanty, A. Khanna, J.J. Rodrigues, D. Gupta, V.H.C. De Albuquerque, Adapting weather conditions based IoT enabled smart irrigation technique in precision agriculture mechanisms, Neural. Comput. Appl. 31 (2019) 277–292.

[31] G. Suciu, H. Ijaz, I. Zatreanu, A.M. Drăgulinescu, Real time analysis of weather parameters and smart agriculture using IoT, in: Future Access Enablers for Ubiquitous and Intelligent Infrastructures: 4th EAI International Conference, FABULOUS 2019, Sofia, Bulgaria, March 28–29, 2019, Proceedings 283, Springer International Publishing, 2019, pp. 181–194.

[32] T.W. Ayele, R. Mehta, Real time temperature prediction using IoT, in: 2018 Second International Conference on Inventive Communication and Computational Technologies (ICICCT), IEEE, 2018, pp. 1114–1117.

[33] N. Golubovic, R. Wolski, C. Krintz, M. Mock, Improving the accuracy of outdoor temperature prediction by iot devices, in: 2019 IEEE International Congress on Internet of Things (ICIOT), IEEE, 2019, pp. 117–124.

[34] Y. Ma, J. Jin, Q. Huang, F. Dan, Data preprocessing of agricultural IoT based on time series analysis, in: Intelligent Computing Theories and Application: 14th International Conference, ICIC 2018, Wuhan, China, August 15–18, 2018, Proceedings, Part I 14, Springer International Publishing, 2018, pp. 219–230.

[35] R. Kumar, P. Kumar, Y. Kumar, Time series data prediction using IoT and machine learning technique, Procedia Comput. Sci. 167 (2020) 373–381.

[36] R. Ab Rahman, U.R.A. Hashim, S. Ahmad, IoT based temperature and humidity monitoring framework, Bull. Electr. Eng. Inform. 9 (1) (2020) 229–237.

[37] G. Stamatescu, C. Drăgana, I. Stamatescu, L. Ichim, D. Popescu, IOT-enabled distributed data processing for precision agriculture, in: 2019 27th Mediterranean Conference on Control and Automation (MED), IEEE, 2019, pp. 286–291.

[38] R. Lomte, G. Patil, C. Patil, N. Sawant, S. Mane, Cropping Pattern Based on Weather Conditions and Soil Composition, 2022. *Available at SSRN* 4040958.

[39] S. Aygün, E.O. Güneş, M.A. Subaşı, S. Alkan, Sensor fusion for IoT-based intelligent agriculture system, in: 2019 8th International Conference on Agro-Geoinformatics (Agro-Geoinformatics), IEEE, 2019, pp. 1–5.

[40] M. Waleed, T.W. Um, T. Kamal, S.M. Usman, Classification of agriculture farm machinery using machine learning and internet of things, Symmetry 13 (3) (2021) 403.

[41] W. Cai, R. Wei, L. Xu, X. Ding, A method for modelling greenhouse temperature using gradient boost decision tree, Inf. Process. Agric. 9 (3) (2022) 343–354.

[42] T.S.T. Bhavani, S. Begum, Agriculture productivity enhancement system using IOT, Int. J. Theor. Appl. Mech. 12 (3) (2017) 543–554.

[43] C. Dewi, R.C. Chen, Decision making based on IoT data collection for precision agriculture, Intell. Inf. Database Syst.: Recent Develop. 11 (2020) 31–42.

[44] F. Edwards-Murphy, M. Magno, P.M. Whelan, J. O'Halloran, E.M. Popovici, B+ WSN: smart beehive with preliminary decision tree analysis for agriculture and honey bee health monitoring, Comput. Electron. Agric. 124 (2016) 211–219.

[45] O. Unold, M. Nikodem, M. Piasecki, K. Szyc, H. Maciejewski, M. Bawiec, M. Zdunek, IoT-based cow health monitoring system, in: Computational Science–ICCS 2020: 20th International Conference, Amsterdam, the Netherlands, June 3–5, 2020, Proceedings, Part V, Springer International Publishing, Cham, 2020, pp. 344–356.

[46] A. Ikram, W. Aslam, R.H.H. Aziz, F. Noor, G.A. Mallah, S. Ikram, I. Ullah, Crop yield maximization using an IoT-based smart decision, J. Sensors (2022) 2022.

[47] S. Turber, C. Smiela, A business model type for the IoT, in: Anais da 26th European Conference on Information Systems (ECIS), Tel Aviv University, Tel Aviv, 2014.

[48] I. Lee, The internet of things for enterprises: an ecosystem, architecture, and IoT service business model, Internet of Things 7 (2019) 100078.

[49] M. Palmaccio, G. Dicuonzo, Z.S. Belyaeva, The internet of things and corporate business models: a systematic literature review, J. Bus. Res. 131 (2021) 610–618.

[50] C. Metallo, R. Agrifoglio, F. Schiavone, J. Mueller, Understanding business model in the internet of things industry, Technol. Forecast. Soc. Change 136 (2018) 298–306.

[51] A. Mishra, M. Mohapatro, Real-time RFID-based item tracking using IoT & efficient inventory management using machine learning, in: 2020 IEEE 4th Conference on Information & Communication Technology (CICT), IEEE, 2020, pp. 1–6.

[52] A. El Jaouhari, Z. Alhilali, J. Arif, S. Fellaki, M. Amejwal, K. Azzouz, Demand forecasting application with regression and IOT based inventory management system: a case study of a semiconductor manufacturing company, Int. J. Eng. Res. Afr. 60 (2022) 189–210. Trans Tech Publications Ltd.

[53] M. Giordano, R. Fischer, M. Crabolu, G. Bellusci, M. Magno, SmartTag: an ultra low power asset tracking and usage analysis IoT device with embedded ML capabilities, in: 2021 IEEE Sensors Applications Symposium (SAS), IEEE, 2021, pp. 1–6.

[54] E. Cakan, A. Şahin, M. Nakip, V. Rodoplu, Multi-layer perceptron decomposition architecture for mobile IoT indoor positioning, in: 2021 IEEE 7th World Forum on Internet of Things (WF-IoT), IEEE, 2021, pp. 253–257.

[55] C.K.M. Lee, C.M. Ip, T. Park, S.Y. Chung, A bluetooth location-based indoor positioning system for asset tracking in warehouse, in: 2019 IEEE International Conference on Industrial Engineering and Engineering Management (IEEM), IEEE, 2019, pp. 1408–1412.

[56] R. Wang, C. Yu, J. Wang, Construction of supply chain financial risk management mode based on internet of things, IEEE Access 7 (2019) 110323–110332.

[57] A. Kanawaday, A. Sane, Machine learning for predictive maintenance of industrial machines using IoT sensor data, in: 2017 8th IEEE International Conference on Software Engineering and Service Science (ICSESS), IEEE, 2017, pp. 87–90.

[58] P. Killeen, B. Ding, I. Kiringa, T. Yeap, IoT-based predictive maintenance for fleet management, Procedia Comput. Sci. 151 (2019) 607–613.

[59] D. Jung, Z. Zhang, M. Winslett, Vibration analysis for iot enabled predictive maintenance, in: 2017 IEEE 33rd International Conference on Data Engineering (ICDE), IEEE, 2017, pp. 1271–1282.

[60] D.T.N. Hashem, The reality of internet of things (Iot) in creating a data-driven marketing opportunity: mediating role of customer relationship management (Crm), J. Theor. Appl. Inf. Technol. 99 (2) (2021).

[61] Z. Li, H. Guo, W.M. Wang, Y. Guan, A.V. Barenji, G.Q. Huang, K.S. McFall, X. Chen, A blockchain and automl approach for open and automated customer service, IEEE Trans. Industr. Inform. 15 (6) (2019) 3642–3651.

[62] I.A. Zualkernan, M. Pasquier, S. Shahriar, M. Towheed, S. Sujith, Using BLE beacons and machine learning for personalized customer experience in smart Cafés, in: 2020 International Conference on Electronics, Information, and Communication (ICEIC), IEEE, 2020, pp. 1–6.

[63] R. Marques, W. de Paula, G.N. Ferreira, F. Armellini, J. Dungen, L. Antonio, de Santa-Eulalia., Exploring the application of IoT in the service station business, IFAC-PapersOnLine 54 (1) (2021) 402–407.

Further reading

[64] B.K. Chae, The evolution of the internet of things (IoT): a computational text analysis, Telecomm. Policy 43 (10) (2019) 101848.

[65] C. Krintz, R. Wolski, N. Golubovic, F. Bakir, Estimating outdoor temperature from CPU temperature for IOT applications in agriculture, in: Proceedings of the 8th International Conference on the Internet of Things, 2018, pp. 1–8.

About the authors

Pan Zheng is currently a Senior Lecturer in Information Systems at the University of Canterbury. His research pursuits encompass the applications of artificial intelligence in healthcare, medicine, and various interdisciplinary domains, as well as a focus on business analytics and data science.

Bee Theng Lau currently holds the position of Professor in ICT and the Associate Dean of Research and Development in the Faculty of Engineering, Computing and Science at Swinburne University of Technology Sarawak. She has been actively contributing to her research areas on assistive technologies for special children, facial expression recognition-based communication, real-time behaviour recognition, smart technologies for the visually impaired, creative art therapies for Autism, STEM education, IoT, etc.

CHAPTER EIGHT

Internet of things challenges and future scope for enhanced living environments

Jie Liu[a] (iD), Hanyang Hu[b], Weiguo Xu[c], and Dan Luo[d] (iD)

[a]School of Digital Media and Design Arts, Beijing University of Posts and Telecommunications, Beijing, China
[b]School of Architecture, Tsinghua University, Beijing, China
[c]Institute of Future Human Habitat, Shenzhen International Graduate School, Tsinghua University, Beijing, China
[d]School of Architecture, University of Queensland, St Lucia, QLD, Australia

Contents

Abstract

The Internet of Things (IoT) has the potential to significantly ameliorate living environments, notably by integrating smart architecture and smart cities. To actualize this potential, however, addressing a range of technical, social, and economic challenges is imperative. This study systematically explores these challenges and delineates the future prospects of IoT in augmenting living environments. The technical obstacles, including issues of data privacy and security, network and device compatibility, and interoperability and standardization, are examined in conjunction with social challenges such as user acceptance and adoption, accessibility, and inclusiveness, as well as ethical

Advances in Computers, Volume 133
ISSN 0065-2458
https://doi.org/10.1016/bs.adcom.2023.10.007

and legal considerations. The economic challenges, such as cost and funding, return on investment, and implementation and deployment processes, are also scrutinized. This work underscores the importance of smart architecture and smart cities in the context of IoT's future scope. It advocates for incorporating IoT within an intelligent ecosystem, which includes technical elements like data collection and integration, advanced data analytics, and real-time monitoring, as well as social and economic aspects like enhanced user experience, increased efficiency, and improved security. The insights presented in this work are of immense value to researchers, practitioners, and policymakers striving to devise and implement robust, inclusive, and ethical IoT solutions. Hence, this study stands as an essential resource, shedding light on the journey towards a more interconnected, efficient, and enhanced living environment.

1. Introduction

The Internet of Things (IoT) has been established as a technology with the rapid evolution and transformative potential in the 21st century. IoT encompasses the interconnectivity of various devices, such as household appliances, wearable technology, and embedded systems, facilitated by the combination of sensors, communication technologies, and embedded systems. This interconnectivity has created a network of devices that can collect, transmit, and exchange data in real-time, thereby increasing efficiency and productivity.

One of the most notable examples of IoT's impact can be seen in the field of smart homes. With the integration of IoT technology, home appliances, such as refrigerators, washing machines, and lighting systems, can be remotely controlled and monitored through a single interface, such as a smartphone app. Homeowners can monitor and control their home's energy consumption, reducing waste and saving on energy bills. Additionally, smart homes equipped with IoT technology can respond to occupants' needs, such as adjusting room temperature [1], lighting [2], and music, based on their preferences and routines. This leads to increased comfort and convenience for homeowners. Another example of the impact of IoT can be seen in the healthcare field. IoT-enabled medical devices, such as wearable fitness trackers and smart pills, have revolutionized how medical professionals monitor patients' health [3] and administer treatments [4]. For instance, wearable fitness trackers can collect and transmit data on physical activity, heart rate, and sleep patterns to healthcare providers, who can use this information to develop customized health and wellness plans for patients. Moreover, IoT has also had a significant impact on the transportation sector through the advancement of smart cities and connected vehicles

[5]. Integrating sensors and communication technologies in transportation systems have enabled the development of smart traffic management systems [6] that can improve traffic flow, reduce congestion, and enhance road safety.

In recent years, integrating the Internet of Things (IoT) with various domains, including architecture, has gained significant attention from researchers and practitioners. One example of this integration is the integration of IoT with smart architecture and smart cities [7], which aims to create intelligent and sustainable environments that can improve the quality of life for residents. This integration is driven by the goal of creating a bright and connected ecosystem where devices can work together harmoniously to achieve a common goal. Several studies have demonstrated the potential of this integration in practice. For instance, researchers demonstrated that integrating IoT with smart architecture could improve building performance and energy efficiency [8]. By using sensors and actuators to monitor and control various building systems, such as heating, ventilation, and air conditioning, the researchers were able to reduce energy consumption by as much as 15% [8]. Similarly, the development of smart cities is another example of the integration of IoT with architecture. Smart cities aim to create sustainable and livable urban environments by integrating IoT technologies, such as connected lighting systems [9], smart waste management [10], and smart transportation systems [11]. One example of a smart city is Amsterdam, which has implemented a smart lighting system that reduces energy consumption and enhances road safety [9]. This system uses IoT technologies, including sensors and communication technologies, to automatically adjust the brightness of streetlights based on the level of ambient light and the presence of pedestrians and vehicles.

Through continued research and development, the potential of this integration to drive innovation and progress in the field of architecture and smart cities is enormous. Despite the growing body of research on IoT and its potential benefits, much still needs to be done to fully realize the potential of IoT in enhancing living environments. This study aims to delve into IoT's challenges and future scope in enhancing living environments, emphasizing the significance of constructing an intelligent IoT ecosystem that encompasses smart products, smart homes, smart architecture, and smart cities. This ecosystem can enhance the intelligence of IoT, providing new opportunities for improving the quality of life in our living environments. However, there is limited research on constructing such an ecosystem and its potential benefits. A comprehensive overview of the current state of research in this field will be presented, along with an examination of the technical, social, and economic challenges posed by IoT integration in architecture. Possible solutions to these challenges will also be explored in this work, highlighting IoT's promising potential to revolutionize how we live and work.

The research methodology for this work will be based on a comprehensive literature review. The review will include a systematic search of relevant academic journals, conference proceedings, and other reputable sources to gather information on the current state of the field and the existing body of knowledge on integrating IoT with architecture and smart cities. The data collected from the literature review will be analyzed and synthesized to identify the challenges and opportunities associated with IoT integration in architecture and smart cities. The review will also identify gaps in the current research and highlight areas for further investigation. This systematic and in-depth review of the existing literature will provide a comprehensive understanding of the field and serve as the foundation for future research direction in this area.

The structure and content of this document will be organized into several sections. This section provides a comprehensive overview of the topic, the purpose and scope of the study, and the research methodology. This section will set the foundation for the rest of the document. The following sections of the document will build upon the foundation laid out in Section 1, starting with a detailed analysis of the technical, social, and economic challenges associated with IoT integration in architecture and smart cities in Section 2. In Section 3, the study will explore the potential of IoT to enhance living environments and the future scope of IoT in smart architecture and smart cities. Section 4 will delve into the concept of an intelligent IoT ecosystem, including its technical and social-economic aspects. Finally, this document will summarize its key findings and implications in Section 5 and conclude with further research and development recommendations.

2. IoT challenges in smart architecture and smart cities

Incorporating the Internet of Things (IoT) across diverse sectors and everyday life has been acknowledged as a significant driving force behind technological innovation and advancement. However, the infusion of IoT into architecture and smart cities encounters a spectrum of technical, social, and economic challenges that should be surmounted to harness its capabilities fully. This section is dedicated to an exhaustive analysis of the challenges of integrating IoT into architecture and smart cities. This section aims to equip readers with a holistic understanding of the technical, social, and economic difficulties faced during integrating IoT into architecture and smart cities. Moreover, it aims to offer evidence-based perspectives that can guide future research and development initiatives in this realm. Each challenge will be thoroughly examined, supported by pertinent cases, examples, and references from credible academic and industry sources.

2.1 Technical challenges

Implementing IoT technology in architecture and smart cities requires the resolution of various technical challenges that impact the systems' reliability, efficiency, and security. The following are some critical technical challenges that must be addressed to achieve the desired outcomes of IoT integration.

2.1.1 Data privacy and security

Data privacy and security are key concerns in smart architecture and smart cities, as the implementation of IoT technology requires the processing and storage of large amounts of sensitive information, such as personally identifiable information, financial information, health information, location data, behavioral data, and infrastructure and security information [12,13]. This information is vulnerable to attacks from cybercriminals, who can use it for malicious purposes such as identity theft, financial fraud, and the spread of malware. To mitigate these risks, it is essential to have effective measures in place to protect the privacy and security of the data generated and collected by IoT systems.

Multiple approaches to addressing the data privacy and security challenge are using encryption technologies, such as Quantum Key Distribution (QKD), Homomorphic Encryption, Secret Sharing Schemes, Elliptic Curve Cryptography (ECC), Format-Preserving Encryption (FPE), Authenticated Encryption, Advanced Encryption Standard (AES), Lightweight Cryptography, Post-Quantum Cryptography, Public Key Infrastructure (PKI). Each technology has its strengths and weaknesses. Based on the last five years of research literature, Table 1 details and analyzes the ten revolutionary encryption technologies.

Another approach to addressing the challenge of ensuring data privacy and security in the context of the smart architecture and smart cities is through the utilization of secure communication protocols. These protocols, such as Secure Socket Layer (SSL) and Transport Layer Security (TLS),

Table 1 Pros and cons of ten revolutionary encryption technologies.

Encryption technologies	Definition	Pros	Cons
Quantum Key Distribution (QKD) [14–16]	A cryptographic technique that enables secure communication by transmitting encryption keys over a quantum communication channel.	Unbreakable security, secure communication over long distances	High cost, technical complexity
Homomorphic Encryption [17–19]	A type of encryption technology that allows computations to be performed directly on encrypted data, without requiring the data to be decrypted first.	Strong security, preserves privacy, no need to decrypt data	Slow processing, computationally intensive, difficult to implement in real-world applications
Secret Sharing Schemes [20,21]	A cryptographic method used to secure sensitive data by dividing the secret into multiple parts, called shares, and distributing the shares among multiple participants.	Flexible implementation, improved data privacy	Limited security, risk of data tampering, complexity
Elliptic Curve Cryptography (ECC) [22,23]	A public-key cryptography system based on the algebraic structure of elliptic curves over finite fields.	Strong security, low computational overhead	Implementation difficulties, risk of quantum attacks
Format-Preserving Encryption (FPE) [24,25]	A cryptographic technique enables the encryption of data while preserving the original format, making it easier to process and analyze the encrypted data.	Flexible implementation, improved data privacy	Limited security, risk of data tampering
Authenticated Encryption [26,27]	A cryptographic technique that provides both confidentiality and integrity protection for the data being transmitted or stored.	Improved security, reduced risk of data tampering	Complex implementation, increased processing overhead
Advanced Encryption Standard (AES) [28,29]	A widely adopted symmetric-key encryption standard, offering high-speed encryption and decryption.	Widely adopted, high-speed processing	Vulnerability to key attacks, limited security
Lightweight Cryptography [30–32]	Cryptographic algorithms and protocols that are designed to be efficient in terms of computation and memory requirements.	Efficient implementation, low computational overhead	Limited security, potential for implementation errors
Post-Quantum Cryptography [33]	A term used to describe cryptographic algorithms that are designed to be secure against attacks from quantum computers.	Future-proof security, resistance to quantum attacks	Slow processing, limited implementation options
Public Key Infrastructure (PKI) [34,35]	A technology provides a secure framework for digital certificates, enabling secure communication and authentication in the IoT.	Strong security, flexible implementation	Complex management, high overhead

serve as a secure means of transmitting data between devices and networks, thereby mitigating the risks associated with eavesdropping and tampering. In recent years, various innovative secure protocols for the Internet of Things (IoT) have emerged, each offering specific advantages and disadvantages. Table 2 provides an overview of the ten most proposed IoT secure protocols in the last five years, detailing their definitions, advantages, and limitations.

In addition to encryption and secure protocols, there is also a growing interest in the use of blockchain technology to address the data privacy and security challenge in smart architecture and smart cities. This is due to its secure and decentralized nature that enables data to be stored and managed in a network of nodes. The decentralized structure of blockchain technology makes it an attractive option for IoT systems that require a high level of security and privacy. It ensures that data stored on the blockchain network is accurate and tamper-proof, making it difficult for malicious actors to access or alter the data without detection. This structure is achieved through the use of encryption and secure protocols, such as public key cryptography [57,58], consensus algorithms [30,59], and digital signatures [60]. One example of using blockchain technology in smart cities is implementing a blockchain-based platform for tracking and managing waste management [61,62]. This platform utilizes blockchain technology to securely store and track data related to waste management, including waste collection, transportation, and disposal. The decentralized structure of blockchain helps to ensure that the data stored on the platform is accurate and tamper-proof. Another example is using blockchain technology for secure data management in smart grids [63,64]. In this application, blockchain is used to securely store data related to energy generation, distribution, and consumption [65]. This allows for a more efficient and secure energy management system, as data related to energy production and consumption is stored in a decentralized and tamper-proof manner. However, while blockchain technology provides several benefits for data privacy and security, some challenges must be addressed. One of the main challenges is scalability [30,66], as the decentralized nature of blockchain can result in slow transaction processing times. This can be addressed through scalable consensus algorithms, such as Proof of Stake (PoS) [67,68] and Delegated Proof of Stake (DPoS) [69,70]. Additionally, there are concerns about the energy consumption [71,72] associated with blockchain, as the process of creating new blocks requires a significant amount of computing power. This can be mitigated through energy-efficient consensus algorithms, such as Proof of Activity (PoA) [73] and Proof of Capacity (PoC) [74,75].

Table 2 Pros and cons of ten most proposed IoT secure protocols.

Secure communication protocols	Definition	Pros	Cons
Secure Socket Layer (SSL) [36–38]	SSL is a cryptographic protocol that provides secure communication over the internet by establishing an encrypted connection between a client and a server.	Confidentiality, authentication, integrity, widely adopted.	Performance overhead, vulnerability to attacks, certificate management, outdated.
Transport Layer Security (TLS) [39,40]	TLS is a widely used security protocol that provides secure communication between devices and networks in the IoT. It uses public key infrastructure (PKI) to establish secure communication channels, and it provides message confidentiality, integrity, and authenticity.	Widely adopted, and secure.	Complex implementation, and high overhead.
DTLS (Datagram Transport Layer Security) [41,42]	DTLS is a security protocol designed to provide confidentiality, authenticity and data integrity for datagram-based applications, such as those using the User Datagram Protocol (UDP).	lightweight, efficient.	limited security compared to TLS.
MQTT (Message Queuing Telemetry Transport) [43,44]	MQTT is a lightweight messaging protocol that is widely used for IoT communication. Its key features include low overhead, publish/subscribe communication, and secure data transmission through the use of SSL/TLS encryption.	Lightweight, low overhead, and easy to implement.	Limited security features, such as the lack of message authenticity and confidentiality.
CoAP (Constrained Application Protocol) [41,45,46]	CoAP is a specialized web transfer protocol for use with constrained nodes and networks in the IoT. It provides a simple and secure way of transmitting data between devices and networks, and it uses DTLS (Datagram Transport Layer Security) for encryption.	Low overhead, simple, and secure.	Limited feature set, as compared to other protocols.

Protocol	Description	Advantages	Limitations
Zigbee [47–49]	Zigbee is a low-power, low-data-rate wireless communications protocol designed for use in IoT devices and networks. It uses AES (Advanced Encryption Standard) for encryption, providing secure data transmission between devices and networks.	Low power consumption, low overhead, and widely supported.	Limited security features, and slow data transmission speed.
LWM2M (Lightweight Machine-to-Machine) [50–52]	LWM2M is a secure communication protocol designed for M2M (machine-to-machine) communication in the IoT. It provides secure data transmission through the use of DTLS encryption, and it supports various communication modes, including server-initiated and client-initiated communication.	Secure, scalable, and flexible.	Limited adoption, and high overhead.
6LoWPAN (IPv6 over Low-Power Wireless Personal Area Networks) [53,54]	6LoWPAN is a communication protocol that provides secure data transmission between devices and networks in low-power IoT networks. It uses AES encryption to ensure secure data transmission, and it supports various network topologies, including star and mesh networks.	Secure, low power consumption, and scalable.	Limited adoption, and high overhead.
OSCOAP (Object Security for Constrained RESTful Environments) [45]	OSCOAP is a secure communication protocol designed for use in IoT networks. It provides secure data transmission through the use of AES encryption, and it supports both server-initiated and client-initiated communication.	Secure, simple, and flexible.	Limited adoption, and high overhead.
AMQP (Advanced Message Queuing Protocol) [55,56]	AMQP is a protocol provides secure communication for message-oriented middleware and is widely used in financial and e-commerce industries.	Secure communication for sensitive data.	Not as widely adopted as other protocols.

To further enhance the privacy and security of data generated and collected by IoT systems, it is important to have effective data management policies [66,76] in place. These policies should outline the measures that must be taken to protect the privacy and security of the data, including the use of encryption technologies, secure protocols, and data access controls. The policies should also specify the measures that must be taken in the event of a data breach, including incident response procedures [77], data backup [78] and recovery strategies [79], and the reporting of incidents [80,81] to relevant authorities.

2.1.2 Network and device compatibility

Integrating multiple devices and networks in IoT systems can result in compatibility issues [82–84] that can negatively impact the functionality and performance of the system. The challenge lies in ensuring that all devices and networks are compatible and can communicate effectively. For instance, in an IoT system, different devices may have different communication technologies, such as Wi-Fi, Zigbee, and Z-Wave, and operating systems, such as Android and iOS. This heterogeneity of devices and networks makes it difficult to ensure that they are all compatible and can communicate with each other effectively. This can lead to a number of issues, such as low system performance [85,86], increased latency [87], and high maintenance costs [88,89]. Furthermore, IoT devices are often manufactured by different vendors, and each vendor may have its proprietary protocols and standards. This can further exacerbate the challenge of ensuring network and device compatibility, as it becomes difficult to ensure that devices from different vendors can work together effectively.

In the context of the smart architecture and smart cities, this is particularly important as it ensures that building systems, such as lighting, heating, and cooling, and city systems, such as traffic management and waste management, can be effectively integrated and controlled. For instance, in a smart building, it is critical that the lighting system is compatible with the heating and cooling system, such that the lighting system can be turned off automatically when the heating and cooling system is not in use. Without compatibility, these systems may work in isolation, reducing the overall efficiency of the building and increasing energy costs. This requires the implementation of standard protocols and technologies, such as MQTT, CoAP, and 6LoWPAN, which allow for seamless data exchange between devices and networks. Another challenge is the integration of legacy devices and networks with new IoT systems. Legacy devices may not be compatible with the latest IoT technologies and may require modification or replacement in order to be integrated into the system. This can be a costly and time-consuming process, and it may also impact the user experience. In order to overcome these compatibility challenges, IoT systems must be designed with compatibility and interoperability in mind. This requires the use of standard technologies and protocols, as well as the development of integrated solutions that can seamlessly connect different devices and networks. Additionally, IoT systems should be designed to be flexible and scalable [90,91], so that they can accommodate future changes and updates without disrupting the overall functionality of the system. One example of a technology that has been developed to address compatibility challenges in IoT systems is the Open Connectivity Foundation (OCF) [92,93], a standard that enables interoperability between devices and networks in the IoT ecosystem. OCF provides a common communication platform and a set of guidelines for IoT device and application developers to follow, which helps to ensure that all devices and networks are compatible and can communicate effectively with each other. Another example is the AllSeen Alliance [94], a cross-industry consortium that has developed an open-source framework for IoT device and application development. The framework provides a common language for devices and applications to communicate with each other, which helps to ensure that all devices and networks are compatible and can communicate effectively with each other. Some numerous other technologies and solutions have been developed to address compatibility challenges in IoT systems, and research in this area continues to advance. The key is to ensure that all devices and networks are compatible and can communicate effectively with each other, in order to enable the full potential of IoT systems to be realized.

2.1.3 Interoperability and standardization

Even though multiple devices from different manufacturers and vendors in IoT systems could be connected to exchange messages, it still has interoperability issues [95] that can negatively impact the functionality and performance of the system. This challenge is rooted in the lack of established industry standards and protocols that ensure interoperability between devices and networks. Establishing industry standards and protocols to ensure interoperability between different devices and networks is considered one of the significant technical challenges of IoT. This challenge is particularly relevant in smart architecture and smart cities where large-scale integration of IoT systems is needed to provide a seamless user experience and support smart living environments.

In smart architecture, the integration of multiple devices and systems from different manufacturers and vendors often results in interoperability issues that impact the overall functionality of the system. For example, a smart home system that includes devices from multiple vendors may not be able to integrate seamlessly, leading to a lack of data exchange and interoperability. In such cases, the end-user may have to switch between different apps to control different devices, leading to a fragmented user experience. To overcome this challenge, industry standards and protocols need to be established to ensure interoperability and compatibility between different devices and systems.

Similarly, in smart cities, integrating multiple IoT systems and devices from different vendors poses a challenge regarding interoperability and standardization. For instance, a smart transportation system that integrates multiple IoT devices and connects to the city's traffic management center requires seamless communication between devices and systems. However, due to the lack of standardization and interoperability, the system could understand the message from different device effectively, leading to inefficiencies in traffic management and control. This can result in increased traffic congestion, longer commute times, and reduced quality of life [96,97] for citizens. Positive examples include the Ann Arbor Connected Vehicle Test Environment (AACVTE) [98] in Michigan, where the city collaborated with the University of Michigan to test connected vehicle technology. The program aimed to improve road safety and reduce traffic congestion by integrating connected vehicles, infrastructure, and data management systems from multiple vendors. The program used standardized communication protocols and data formats to ensure interoperability and seamless data exchange between the various systems, which led to a significant reduction in traffic congestion and improved safety.

To address these challenges, several innovative academic studies have been conducted in the field of smart cities. One such study proposed a multi-layer interoperability framework for smart cities [99,100], which includes physical, data, and semantic layers. This framework aims to ensure the compatibility and interoperability of IoT devices and systems within a smart city environment by standardizing the data exchange between these devices. Another study proposed a cloud-based platform for smart cities [101,102] that incorporates data and system interoperability. The platform allows for integrating multiple IoT devices and systems and enables the central management and control of these devices. This approach has improved the interoperability and standardization of IoT systems in smart cities, resulting in improved efficiency and cost savings. However, several obstacles may still exist in enhancing interoperability and standardization in smart architecture and smart cities. For instance, the lack of standardization in IoT devices and systems, and the absence of regulatory bodies to enforce these standards, may continue to pose a challenge in terms of interoperability. Additionally, the high cost of replacing existing IoT devices and systems with standardized and interoperable ones may hinder the implementation of interoperability and standardization measures in these environments.

2.2 Social challenges

In addition to the technical challenges, integrating IoT into architecture and smart cities also poses various social challenges that impact the adoption and acceptance of the technology. The following are some of the critical social challenges that must be addressed to ensure the successful adoption of IoT technology:

2.2.1 User acceptance and adoption

The widespread adoption of IoT technology depends on users embracing and utilizing the technology. For this to happen, users must be thoroughly informed and educated about the technology and its benefits. However, a significant challenge to achieving widespread adoption is overcoming the resistance to change and building trust and confidence in the technology among users.

Studies have revealed that lack of awareness and education, privacy concerns, and trust in technology are the main reasons for low adoption rates. To address these challenges, researchers have proposed various strategies, including user education and awareness programs [103], transparency and accountability [104] in data collection and use, and the development of security protocols [105] to protect personal data. For example, the study found that providing users with clear information about IoT devices and their capabilities and the measures taken to protect their data significantly increased users' trust and confidence in the technology [106]. Researchers also developed a privacy-enhancing framework for IoT devices that minimizes the collection of personal data and enhances user control over the data collected [107]. These findings highlight the importance of addressing the social challenges related to user acceptance and adoption in order to achieve widespread adoption of IoT technology in smart architecture and smart cities. It is crucial for researchers, policymakers, and industry leaders to work together to develop solutions that address these challenges and foster trust and confidence in the technology.

2.2.2 Accessibility and inclusiveness

In terms of accessibility and inclusiveness in IoT technology, there have been numerous studies that have investigated the barriers and challenges faced by individuals with disabilities and special needs. For instance, a World Health Organization (WHO) study found that nearly 1 billion people with disabilities and elders are often excluded from the benefits of new technologies, including IoT, due to a lack of accessibility and inclusiveness [108]. Various initiatives have been launched to address these issues to promote accessibility and inclusiveness in IoT technology. For example, the National Accessibility Programme in South Africa aims to develop accessible and inclusive IoT solutions for people with disabilities and special needs [109]. This project involves collaboration between universities, research institutions, and industry partners to develop accessible and inclusive IoT solutions that meet the needs of people with disabilities and special needs.

In addition, various standards and guidelines have been developed to ensure that IoT technology is accessible and inclusive. For instance, the Web Content Accessibility Guidelines (WCAG) provide guidelines for making web content more accessible to people with disabilities and are often used as a reference for developing accessible IoT solutions [110,111].

Research has shown that addressing the issues of accessibility and inclusiveness in IoT technology can have a positive impact on the adoption and usage of the technology by individuals with disabilities and special needs. For example, a study conducted by the International Association of Accessibility Professionals (IAAP) found that accessible IoT solutions can increase the independence and quality of life for individuals with disabilities, as well as promote social inclusion [112]. Research has shown a significant gap in the accessibility and usability of IoT devices for individuals with disabilities. Study shows less than 1% of IoT devices are fully accessible to visually impaired individuals. This highlights the importance of including accessibility considerations in the design and development of IoT technology.

To maximize the benefits of IoT technology, it must be accessible and inclusive to all users regardless of their background or abilities. This requires ensuring that the technology is usable and accessible to people with disabilities and special needs and does not discriminate against certain groups. Some examples of ensuring accessibility and inclusiveness in IoT technology include designing user interfaces that are compatible with assistive technologies such as screen readers [113], keyboard navigation [114], and force feedback [115], incorporating text-to-speech and speech-to-text functionality for users with visual or mobility impairments, and designing products [116] with consideration for a range of ages, cultures, and language abilities. Other technology, like blockchain, also plays an important role in supporting accessible smart cities [117].

To address these issues, researchers and organizations are exploring various approaches and techniques, such as using universal design principles [118], user-centered design [119], and accessibility testing and evaluation methods [120]. These projects usually involve a multi-disciplinary team of researchers, designers, and accessibility experts. They are focused on creating guidelines and standards for the accessibility of IoT products and services. For example, the Smart Inclusive City aims to develop an inclusive framework for the design and development of IoT technology that addresses the needs of disabilities [121]. Universidad Politécnica de Madrid also presented a system architecture for the interconnection and cooperation of different IoT networks, multimodal interfaces, and service robots for disabilities [122].

2.2.3 Ethical and legal considerations

IoT technology has brought about significant ethical and legal considerations, including data privacy, protection of personal information, and responsible technology use [13,123]. There has been much discussion in academic and legal circles regarding the development of ethical and legal frameworks that can effectively balance the benefits of IoT technology with the privacy and rights of individuals. Studies have shown that data privacy is a major concern for individuals regarding IoT technology. For example, the study found that nearly 80% of individuals were concerned about the privacy of their personal data in the context of IoT technology [124]. Another study conducted by Accenture found that 47% of individuals rejected IoT devices since the data security concerns [124].

In response to these concerns, researchers have been developing new techniques for enhancing the privacy and security of personal data in IoT systems. For instance, encryption [125], secure data storage [126], and access control mechanisms [127] have been proposed as potential solutions for enhancing the privacy and security of personal data in IoT systems. Additionally, researchers have been exploring the use of blockchain technology and other decentralized systems to enhance data privacy and security in IoT systems.

Regarding legal considerations, there have been various efforts to develop frameworks for regulating the use of IoT technology. For example, the European Union has developed the General Data Protection Regulation (GDPR) [128], which aims to enhance the privacy rights of individuals in the context of IoT technology. Additionally, various countries have enacted laws aimed at regulating the use of IoT technology, such as the Electronic Communications Privacy Act (ECPA) [129] in the United States and the Personal Information Protection and Electronic Documents Act (PIPEDA) [130] in Canada. These laws provide guidelines for the collection, use, and storage of personal data collected through IoT devices, as well as rules for ensuring the security and privacy of such data.

In terms of ethical considerations, IoT technology raises various ethical questions, such as the appropriate use of data, the impact of IoT technology on society, and the accountability of organizations that use IoT technology. To address these issues, researchers and industry leaders have called for developing ethical frameworks and codes [131] of conduct that can guide the development and use of IoT technology. For instance, the IEEE Standards Association has developed the IEEE P7000 [132] series of standards, which provides guidelines for the ethical design and development of IoT technology.

2.3 Economic challenges

Integrating Internet of Things (IoT) technology into architecture and smart cities has been identified as a promising solution for enhancing the sustainability, efficiency, and livability of urban environments. However, the implementation and deployment of IoT systems also pose various economic challenges [133] that impact the feasibility and sustainability of these systems. This study will examine three key economic challenges associated with IoT in smart architecture and cities: cost and funding, return on investment and monetization, and implementation and deployment.

2.3.1 Cost and funding

One of the major economic challenges in implementing IoT technology in smart architecture and cities is the high cost associated with the deployment and maintenance of these systems. The cost of IoT systems includes the cost of hardware, software, network infrastructure, and data management and analysis. The cost of hardware and software components is relatively high, and the cost of network infrastructure and data management can also be significant. The challenge lies in securing adequate funding and financing for implementing these systems.

There are several funding options available for the implementation of IoT systems in smart architecture and cities, including government grants and loans, private investment, and public-private partnerships [134]. However, the availability and terms of these funding options can vary widely depending on the country, region, and sector. In addition, there is often a lack of clarity and consistency in the terms and conditions of funding, making it difficult for organizations to secure adequate funding.

One example of a funding initiative aimed at promoting the implementation of IoT systems in smart architecture and cities is the European Commission's Horizon 2020 program, which provides funding for research and innovation in the field of IoT. The program provides funding for projects that address key challenges associated with deploying IoT systems, including cost and funding [135–137].

2.3.2 Return on investment and monetization

Another major economic challenge in implementing IoT systems in smart architecture and cities is the return on investment (ROI) and monetization potential of these systems. The challenge lies in balancing the costs and benefits of the technology and ensuring that the ROI and monetization potential is adequate to support the long-term viability of the systems.

The monetization potential of IoT systems in smart architecture and cities is largely dependent on the value generated by the systems. For example, the value of energy savings generated by smart lighting and building management systems can be monetized through energy savings or carbon credits. The value generated by other IoT systems, such as traffic management systems and environmental monitoring systems, can be monetized through improved efficiency, reduced costs, and increased revenues.

There have been several studies aimed at evaluating the ROI and monetization potential of IoT systems in smart architecture and cities. For example, a study by the European Network of Living Labs found that the ROI of smart lighting systems can be significant, with payback periods ranging from 1 to 3 years. Another study by the World Economic Forum found that the monetization potential of IoT systems in the city of Amsterdam was estimated to be over €1 billion per year.

2.3.3 Implementation and deployment

The implementation and deployment of IoT systems in smart architecture and cities is a complex and challenging process that requires the coordination of multiple stakeholders, including governments, businesses, and communities. The challenge lies in ensuring the effective and efficient deployment of these systems and the integration of these systems into existing urban infrastructure. There are several approaches to the implementation and deployment of IoT systems in smart architecture and cities, including top-down and bottom-up approaches. The top-down approach involves the centralization of decision-making and the deployment of systems by government agencies or large corporations. The bottom-up approach involves the decentralization of decision-making and the deployment of systems by communities or local organizations.

There have been several examples of successful implementation and deployment of IoT systems in smart architecture and cities. For instance, the city of Amsterdam has implemented a number of IoT systems, including traffic management systems, smart lighting systems, and environmental monitoring systems, with the aim of enhancing the sustainability and livability of the city. In addition, the city of Barcelona has implemented a smart city platform that integrates various IoT systems, including traffic management, environmental monitoring, and energy management systems. The platform has been instrumental in improving the efficiency, sustainability, and livability of the city.

3. Future scope of IoT in enhancing living environments

The integration of IoT technology in architecture and cities has the potential to transform the way we live and work, creating more sustainable, efficient, and livable environments. This section will explore the future scope of IoT in enhancing living environments through two main approaches: smart architecture and smart cities.

3.1 Smart architecture

Architecture plays a crucial role in shaping the living environment and determining the physical and aesthetic characteristics of the built environment. With the integration of IoT technology, architecture has the potential to become even more impactful, as it can help to design intelligent living environments that are not only visually appealing but also highly functional and efficient. IoT technology can be integrated into the design of buildings, homes, and other structures, to create intelligent and connected environments that can adapt to changing needs and requirements.

There have been several famous examples of smart architecture in recent years, with innovative and cutting-edge designs that incorporate the latest technologies to improve the functionality and sustainability of buildings. The Edge in Amsterdam with over 28,000 sensors that collect and analyze data to optimize energy usage, lighting, temperature, and other factors to provide a comfortable and efficient workspace [138]. The Shanghai Tower [139] in Shanghai has a regenerative braking system for the elevators, a double-skin façade that helps to reduce energy consumption, and a rainwater harvesting system that provides up to 20% of the building's water needs.

As a platform of IoT devices, smart architecture incubates multiple application scenarios, such as smart homes, intelligent buildings, and connected infrastructure. Smart homes, for example, can be equipped with IoT technology to enable remote monitoring and control of various functions, such as lighting, heating, and security. This can result in energy savings, improved comfort and security, and reduced costs. On the other hand, intelligent buildings can use IoT technology to optimize building performance, reduce energy consumption, and enhance the comfort and safety of occupants. Connected infrastructure, such as smart bridges and roads, can use IoT technology to monitor and improve their performance, ensuring the safety and efficiency of the built environment.

In the future, integrating the IoT devices and architecture can potentially create a more intelligent living environment. A seamless ambient intelligent living environment can be realized by connecting various service robots such as cooking robots, robot arms, virtual gym instructor mirrors, automated organizing robots, cleaning robots, delivery robots, and virtual health control systems. The architecture platform is a central hub that effectively connects all the devices, allowing for a coordinated system response. For instance, after completing a gym exercise at home using a smart mirror with an augmented tutor, the ambient environment can automatically adjust the temperature and humidity to a more comfortable level based on the wearable device data while the lights dim and the shower starts. The health system can record the energy consumed during the exercise and communicate with the cooking robot and smart fridge to suggest a healthy recipe for the user. In the case of missing ingredients, the delivery robot can be activated to supply the necessary items, subject to user approval. Moreover, the virtual health control system can communicate with the self-driving car and redirect the vehicle to the hospital in case of potential health issues. The potential benefits of IoT-enabled architecture are manifold, offering intelligent and convenient living environments with improved quality of life.

Also, smart architecture can detect and report unusual infrastructure conditions immediately. For example, sensors installed on the building's foundation can detect changes in the building's orientation, indicating potential structural damage. Similarly, sensors installed on pipes and electrical systems can detect leaks, cracks, and other issues, enabling timely repairs. These sensors can send data to a central control system, which uses data analytics and machine learning algorithms to analyze the data and identify potential issues. Once the damage is detected, repair robots can be deployed to perform repairs. These robots can be programmed to perform a wide range of tasks, from patching leaks and fixing cracks to replacing damaged components. Repair robots can also access hard-to-reach areas, such as high ceilings and narrow spaces, enabling repairs that would be difficult or impossible for human workers. Smart architecture cooperation with robots can reduce the need for manual labor, reduce the risk of human error, and improve repair quality and efficiency.

3.2 Smart cities

A smart city is a city that uses advanced technology and data analytics to improve the quality of life, sustainability, and efficiency of its urban environment [140]. The components of a smart city include connected infrastructure, smart buildings, and data analytics and management systems. These components work together to create a more livable, sustainable, and efficient city.

Examples of smart cities include Amsterdam, Singapore, and Barcelona. These cities have implemented IoT technology to improve various aspects of urban life, such as traffic management, energy efficiency, and environmental sustainability. The rapid development and penetration of information technology have led to the gradual transformation of traditional cities into smart cities. The United States first proposed the National Information Infrastructure (NII) and Global Information Infrastructure (GII) in 1993 to promote openness and fair competition in the global information communication market [141]. As a representative of large cities in the United States, New York has also transformed into a smart city through the participation of government and citizens, involving themes such as environmental governance, information infrastructure construction, and public transportation [142]. San Francisco, the location of numerous internet companies, places great importance on constructing information infrastructure, energy utilization, intelligent buildings, and intelligent transportation systems. At the same time, San Francisco's policy direction and organization of the technology community support promoting an intelligent society [143]. In recent years, many European countries have made significant progress in building an intelligent society through the Living-in.EU movement, local data platforms, data space for smart communities, local digital twins, and the DIGITAL program of the European Union [144]. Amsterdam is one of the earliest cities in Europe to adopt an intelligent society strategy. In recent years, it has achieved remarkable results in building smart city platforms and promoting projects. Other cities can directly learn Amsterdam's methods for an intelligent society through the platform mechanism, which fully leverages the advantages of government, businesses, research institutions, and citizens on an open platform [145]. Vienna was also one of the first European cities to propose the construction of a smart city. In 2014, the City Council of Vienna officially released the Smart City Vienna framework strategy and has updated it continuously. Vienna's strategy relies on information and communications technologies (ICT) and has network infrastructure construction as its core. It has also been widely applied in energy efficiency, the environment, and public transportation, solving the sustainability issues of the urban environment while being centered on people [146]. The construction of an intelligent society in Barcelona began in 2009, and by 2012, it became a benchmark for smart cities in Europe. The government places great importance on the role of the community. The 22@Barcelona district was the first to promote infrastructure construction, covering all aspects of the city, from network infrastructures and systems, intelligent living, intelligent governance, and innovation clusters, to people's training programs, traffic management [147].

In the future, the seamless integration of different technologies such as IoT, AI, robotics, and big data analytics will create a highly intelligent and interconnected urban system. For example, the city has a highly advanced resource allocation and management system that ensures the optimal use of resources such as energy, water, and waste. The resource allocation system is greatly flexible and can adjust to the city's changing demands in real time, ensuring that resources are always available where and when needed. Moreover, the smart city's infrastructure has been designed to be exceedingly adaptive and responsive to the needs of its citizens. The city's transportation network utilizes self-driving cars and intelligent traffic systems to provide efficient and safe transportation for the city's residents. With the help of IoT, AI, and big data analytics, the transportation network can optimize the routes and schedules of vehicles, reducing traffic congestion and making the city's transportation system more eco-friendly. The city's highly sophisticated surveillance and security system employs advanced technologies such as facial recognition, video analytics, and sensor networks to monitor the city's public spaces and identify potential threats. The city's emergency response system is extremely efficient and can respond quickly. The city also boasts a highly advanced healthcare system, where precision medicine and telemedicine technologies provide its citizens personalized and remote healthcare services. The city's hospitals and clinics are equipped with advanced medical technologies, such as robotics and AI-powered diagnostic tools, which provide highly accurate and timely diagnoses and treatment.

4. Enhancing IoT with intelligence ecosystems

As the proliferation of smart devices and systems continues and the complexity of IoT systems increases in the future living environment, security, privacy, and interoperability challenges become increasingly prevalent and can negatively impact the user experience. Hence, it is imperative to construct an intelligent ecosystem environment that prioritizes the needs of individuals.

4.1 Intelligent IoT ecosystems

An Intelligent IoT Ecosystem can be defined as a multi-layered system that integrates smart devices and technology to create a seamless and interconnected environment. According to the traditional ecosystem, the intelligent IoT ecosystem has two characteristics.

4.1.1 Multidimensional classification characteristics

A healthy and stable ecosystem often comprises multiple species, diverse classification methods, and various hierarchical relationships to form a multidimensional energy network that facilitates efficient energy circulation. Each major ecosystem can be divided into multiple smaller ecosystems, but different ecosystems do not exist independently and can also influence and interact. There are many ways to classify ecosystems, which can be divided into multiple subsystems based on functional and target differences,

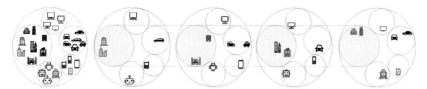

Fig. 1 Different small-scale intelligent ecosystems could be built according to the target functions.

such as the Earth's ecosystem that can be divided into marine ecosystems, forest ecosystems, etc., or by different spatial locations, such as the Amazon ecosystem, the Mediterranean ecosystem, etc.

The construction of an intelligent IoT ecosystem is no different. Different small-scale intelligent ecosystems could be built according to the target functions, such as intelligent residential ecosystems, intelligent transportation ecosystems, and intelligent economic ecosystem. Regionalized ecosystems can also be developed according to a particular location's unique characteristics and demands. For instance, the Manhattan intelligent ecosystem could be designed to address the unique needs and challenges of a dense urban environment. In contrast, the Beijing intelligent ecosystem could be tailored to the requirements of a rapidly growing, technology-driven economy. In this way, intelligent ecosystems can be customized to the needs of specific regions, ensuring that resources are used most efficiently and effectively as possible (Fig. 1).

4.1.2 Fractal characteristics

Ecological systems exhibit fractal properties, with systems at different scales nested within one another. At a global scale, all species and environmental elements on Earth form a large ecological system. In contrast, a single individual organism is composed of various cells and tissues at a local scale. Each hierarchical level is nested within the next, and when a unit stabilizes, it becomes a component of the larger system to which it belongs. Conversely, when a lower-level unit is broken down, its components are integrated into the larger system to which it belongs.

In the same way, intelligent IoT ecosystems can be designed and structured hierarchically, with each layer representing an increasingly complex and sophisticated level of functionality. For instance, an intelligent residential ecosystem could be divided into four layers: smart hardware like sensors and actuators, smart products, smart architecture, and smart cities.

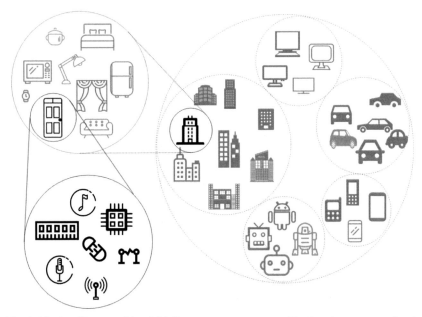

Fig. 2 The intelligent residential IoT ecosystem constructed by four layers: smart hardware like sensors and actuators, smart products, smart architecture, and smart cities.

The integration of these layers creates a dynamic, interconnected ecosystem that can adapt to changing circumstances and optimize the use of resources in real-time (Fig. 2).

With these two characteristics, a well-designed Intelligent IoT Ecosystem would be established. It is possible to systematically and sustainably enhance energy-saving modes and improve user experiences. Specifically, we can analyze from technical, social and economic perspectives how to help the ecosystem operate well.

4.2 Technical aspects of an intelligent IoT ecosystem

To ensure the effective functioning of an intelligent IoT ecosystem, it is essential to focus on three key technical aspects: data collection and integration from multiple sources, advanced data analysis and machine learning algorithms, and real-time monitoring and control of IoT devices.

4.2.1 Data collection and integration from multiple sources

One of the crucial technical aspects for achieving a well-functioning intelligent IoT ecosystem in promoting a sustainable living environment is the data collection and integration from multiple sources. This requires gathering data from various resources, such as different sensors and actuators from different individuals in the ecosystem layers, and integrating this data into a centralized system. This process allows for collecting large amounts of data from multiple sources, enabling comprehensive and accurate analysis of the environment, and supporting decision-making processes that can ultimately improve the overall sustainability and functionality of the ecosystem.

In a smart architecture environment, numerous smart products equipped with sensors and actuators, such as smart lighting, thermostats, and appliances, continuously collect data on usage and behavior. The collected data is then centrally integrated and processed by the smart architecture system to develop insights that inform the optimization of product usage, improve energy efficiency, and enhance overall user experience. The smart architecture system can deduce potential user behaviors and preferences based on past usage patterns, optimize resource utilization in smart homes, such as energy consumption, and provide a more personalized and user-friendly experience. Similarly, in a smart city, data from different smart architectures across the city is centrally processed and integrated, creating a comprehensive view of the data. This facilitates sharing of resources between different smart architectures and provides a more in-depth understanding of the city's behavior and needs, enabling the optimization of city resources such as energy consumption, transportation, and public services.

An intelligent IoT ecosystem's ability to collect and integrate data from various sources is crucial to providing insights into the behavior of IoT devices and the environment in which they operate. Wearable devices, for example, collect data on physical activity, heart rate, and sleep patterns, which can be integrated with data from other sources, such as environmental sensors or medical records. Similarly, data collected from smart homes can be integrated with data from smart transportation systems to provide a more comprehensive view of a user's daily routines, enabling the provision of personalized services and recommendations.

In this context, multiple methods and techniques, such as standardization, data fusion and should be borrowed into the picture. Standardizing data formats and communication protocols, like the Industrial Internet Consortium (IIC) [148], can help ensure that data from different sources can be integrated seamlessly. At the same time, data fusion could help combining data from multiple sources to create a more accurate and comprehensive view of a particular situation. With the growth of big data and the

Intelligent IoT Ecosystem, artificial intelligence (AI) techniques such as convolutional neural networks (CNNs) [149], probabilistic methods, such as Bayesian networks [150] and Markov random fields [151], and distributed data fusion [152] may contribute to identify patterns and relationships in large datasets automatically.

4.2.2 Advanced-data analysis and machine learning algorithms

The increasing deployment of IoT devices will lead to the generation of vast amounts of complex data, which traditional data analysis techniques may struggle to handle. Although different intelligent IoT ecosystem layers could compute the data, utilizing the latest advancements in machine learning and big data analytics is still essential for realizing the full potential of IoT ecosystems. Machine learning and cloud computing algorithms, in particular, offer significant benefits in their ability to analyze large and complex data sets and extract valuable insights from them. Through identifying patterns and relationships within the data, these algorithms can provide valuable insights that can inform decision-making processes and enhance user experience, ultimately improving efficiency and productivity within the ecosystem.

One notable application of machine learning algorithms in the context of IoT is predicting the behavior and preferences of users based on their past usage patterns. By analyzing data from multiple sources, machine learning algorithms such as Long short-term memory (LSTM) neural networks [153], Random Forest [154], Gradient boosting [155], Gaussian mixture models (GMMs) [156], Support vector machines (SVMs) [157], K-nearest neighbors (KNN) [158], Hidden Markov models (HMMs) [159], Collaborative filtering [160] and Matrix factorization [161] can personalize services and optimize resource utilization, ultimately improving the overall user experience. Additionally, machine learning algorithms can be used for anomaly detection and identifying behavior patterns that may indicate potential issues or problems. By identifying and addressing issues before they escalate, these algorithms can maintain uptime and ensure safety within the intelligent IoT ecosystem.

There also some advanced algorithms for fasting the decision-making procedure. The use of edge computing and fog computing has gained prominence in smart city IoT research. Edge computing [162] involves processing data and executing applications closer to the network's edge, where IoT devices are located, to reduce latency, improve response times, and enhance privacy and security. Meanwhile, fog computing [163] is a related concept that focuses on decentralizing computing architecture and reducing cloud computing capabilities to the edge of the network, supporting large-scale IoT applications. Multi-access edge computing (MEC) [164] is a promising approach for developing an intelligent IoT ecosystem, offering a wide range of applications and addressing the requirements of low latency, distributed deployment, location awareness, and mobile support. Edge AI, or distributed neural networks (DNN) [165], provides the possibility of achieving intelligent and seamless interactions between the physical and digital worlds, enhancing the efficiency and security of IoT in the intelligent IoT ecosystem. The cloud-edge-terminal system [166], which adaptively distributes DNN model layers and data samples, can also accelerate the DNN modeling process. Edge cloud [167], being smaller and closer to the device than the core cloud, extends cloud capabilities at the edge by leveraging user- or operator-contributed computes nodes at the edge of the network. Network Function Virtualization (NFV) [168] and Software-Defined Networking (SDN) [169] are promising technologies for managing future networks, offering benefits such as cost savings, resource utilization, and efficient management.

4.2.3 Real-time monitoring and control of IoT devices

The importance of real-time monitoring and control of IoT devices cannot be overstated in the quest to enhance IoT with intelligent ecosystems. With the increasing ubiquity of IoT devices, they have been deployed in various settings such as homes, buildings, and cities. These devices provide a wealth of data that can be analyzed to optimize their operation and improve the user experience. To achieve these benefits, it is crucial to monitor and control the devices in real-time.

Real-time monitoring and control of IoT devices allow people to receive immediate feedback on the status of the devices and their operations. This feedback helps users to gain a better understanding of how the IoT ecosystem is working and how their actions are affecting it. For instance, the use of digital twins [170] of smart architectures and smart cities enables users to visualize the data from IoT devices in real-time, allowing for easy identification of problems and timely resolution. New user interfaces, such as the Microsoft HoloLens mixed reality headset, allow users to interact with IoT devices in real-time, providing a more immersive and intuitive experience. By visualizing and controlling IoT devices in 3D, it becomes easier to understand their functionality and operation.

To ensure that all people, including the elderly and disabled, can participate in intelligent IoT ecosystems, multimodal dialogue systems for the elderly population [171] would be established, the feedback can be in the form of pictures, sound, tangible interfaces, or even olfactory displays depending on the device and user preference.

Real-time monitoring and control of IoT devices also enables the system to respond to changes quickly. For example, if a security sensor detects a potential threat, the system can alert the appropriate authorities and take corrective action to prevent damage or loss. Moreover, with advances in smart materials such as shape memory alloys (SMAs) [172] and piezoelectric materials [173], the living environment can become more interactive. These materials can be integrated with IoT sensors and actuators to create smart structures that respond to changes in their environment in real-time. By adjusting physical properties such as temperature, pressure, and strain, these smart structures can provide an innovative type of smart environment interface.

With real- time monitoring in an active and intelligent living environment, the intelligent IoT ecosystem has great potential to provide critical assistance to minority groups. For instance, if an individual with a disability falls, sensors embedded within the IoT network can detect the fall and immediately alert emergency services or family members. Simultaneously, the intelligent IoT ecosystem can send monitoring data to these parties to inform them of the individual's status. The system can also facilitate efficient transportation for emergency services by communicating with the city's transportation system to ensure a prompt response.

4.3 Social and economic aspects of an intelligent IoT ecosystem

The integration of advanced intelligence into the Internet of Things (IoT) has the potential to revolutionize various aspects of daily life, particularly in terms of enhancing living environments. The social and economic aspects of an intelligent IoT ecosystem are particularly critical in realizing this potential. In this section, we will explore the significance of these aspects and how they can contribute to improved user experience, increased efficiency, and enhanced security and privacy.

4.3.1 Improved user experience and personalized services

An intelligent IoT ecosystem has the potential to revolutionize user experience and provide personalized services. By incorporating AI and machine learning technologies into IoT devices, and following the principles of user-centered design, it is possible to create a more intuitive and responsive system that can anticipate and meet the user's needs.

One of the key benefits of an intelligent IoT ecosystem is its ability to create a more active healthy and thoughtful environment. For instance, elderly individuals often fall asleep while watching TV on the sofa, leading to poor blood circulation and a decline in physical health. Smart products, when integrated with a smart architecture ecosystem, can mitigate this situation. Wearable devices can monitor various aspects of human health, such as body temperature, heart rate, blood pressure, and glucose levels, to determine whether the elderly person has fallen asleep and the quality of their sleep and synchronize the data with the smart architecture. When the smart architecture detects that the elderly person has fallen asleep, it sends a signal to the smart TV to gradually reduce the TV volume, as well as the brightness of the TV and the indoor lighting, until it turns off. Additionally, when the elderly person enters deep sleep, the smart architecture sends a signal to the smart sofa. The sofa slowly adjusts its shape to flatten the backrest until it provides a surface with an average pressure distribution for the elderly person, thereby enhancing their sleep quality. Furthermore, the intelligent building automatically adjusts the indoor environment temperature based on the elderly person's body temperature to prevent the elderly person from catching a cold during the sleep stage.

Another way an intelligent IoT ecosystem can improve the user experience is by facilitating the transfer of personalized information across different intelligent devices. Wearable devices can serve as portable data carriers that can be moved around a physical space, enabling different intelligent building entities to quickly and effectively obtain the same user's personalized information, allowing them to provide more tailored services to the building's occupants. For instance, a smart wristband can record various data types, including the building's commands and other related interaction data. It can then create a behavioral map that can be uploaded to the intelligent building when the user enters a new environment, providing the building with precise information about its preferences and enabling it to offer personalized services.

Moreover, wearable devices often have additional features, such as location tracking and acceleration sensing, that can help the building locate its occupants. For example, a wearable device can regulate the temperature of the indoor environment by tracking the user's heart rate, body temperature, and ambient temperature and determining the range of indoor environmental temperatures the user can tolerate. When the user enters an unfamiliar environment, the wearable device automatically uploads this information to the intelligent building, which then adjusts the indoor environment temperature based on the temperature preferences of all occupants, thereby maximizing the indoor temperature comfort for the user group.

4.3.2 Increased efficiency and cost savings

One potential solution to address the challenges posed by the ever-increasing energy consumption and environmental impact in the intelligent IoT ecosystem is the implementation of advanced data analysis and machine learning algorithms to improve efficiency and reduce costs. As IoT devices proliferate, energy efficiency and long battery life become increasingly crucial. Therefore, researchers are exploring ways to reduce the power consumption of IoT devices, develop energy-efficient protocols and algorithms, and monitor and control energy consumption in intelligent IoT ecosystems.

Wireless sensing networks (WSNs) [174] have emerged as a powerful tool that enables cities to access real-time data through sensors, which can monitor, understand, analyze, and plan cities to improve energy efficiency and environmental health in real-time. Smart Grid [63] technology is another promising solution that enables real-time measurement and monitoring through metering, communication, and cloud computing technologies. It has applications in home and building automation, smart substation, and feeder automation, which can contribute to environmental protection and energy sustainability more effectively. Context-aware systems [175] are another critical area for the development of intelligent systems. They have great potential for practical implementation in various domains, such as ambient intelligence, active and assisted living, IoT, mobile computing, and wireless sensor networks. For instance, a smart home might use a context-aware system to adjust the temperature and lighting based on the time of day, whether anyone is home and other factors. Machine learning techniques have been widely used to improve energy efficiency, which can significantly reduce energy consumption and achieve energy savings and emission reduction. Support vector regression (SVR) [176], artificial neural networks (ANNs) [177], classification and regression tree (CART) [178], and general linear models [179] are commonly used in machine learning applications. Green IoT [180] is a research direction aimed at reducing the environmental impact of IoT systems through the energy-efficient design and renewable energy sources. Energy harvesting [181] is another developing technology that allows IoT devices to be powered by the energy they generate from their environments, such as solar, wind, or kinetic energy.

Another potential solution to increased efficiency and cost savings is to design flexible physical properties for infrastructures and products that can help them adapt to changing needs and prolong their lifespan, leading to long-term cost savings. Take smart architecture as an example. In this context, a smart architecture needs to ensure the flexibility of its various spatial enclosure components and combinations during the design process to adapt to changes in future social lifestyles effectively. Therefore, the building needs to be flexible in spatial layout and function.

Architects can design buildings as a framework that is the building's load-bearing structure that can load various spatial service components, integrating to form a system. The framework has enough flexible shape-changing walls, movable furniture, or small spatial units with sensing and deforming capabilities. We can regard the minimum unit that constitutes smart architecture as having modular characteristics, where shape-changing small units with different functions are combined to form a large building space. The shape and function of this building space are determined by the small units that constitute it, but different building spaces can be combined and suspended on the building's load-bearing frame to form the entire building. The functions within the building space also can change with different demands. As people's lifestyles update, their demands for spatial functions are also changing. Spatial functions are becoming more open and diversified. Therefore, in the future, the functions and nature of space in smart architecture may change to people's needs. The smart architecture of the future will change the functional nature of the building space at any time with the high-speed iteration of users' needs for building space. The functional attributes of the building will become ambiguous, complex, and diverse in this constantly changing environment. Therefore, an intelligent IoT ecosystem requires the building to continuously update its database based on social needs, update the qualitative descriptions of "functional requirements" in the building.

4.3.3 Enhanced security and privacy

An intelligent IoT ecosystem can enhance data security and privacy through the use of advanced technologies and private-friendly interaction design to monitor and control devices and protect sensitive data. From a technological standpoint, Physical Unclonable Functions (PUFs) [182] provide a valuable advantage in securing IoT systems by exploiting inherent physical differences to generate hardware fingerprints. However, researchers are currently working to reduce the False Rejection Rate and False Authentication Rate of PUFs [183]. Arcenegui et al. [184] proposed a secure management scheme for IoT devices that leverages blockchain and PUFs. In this scheme, participants manage their Blockchain Accounts (BCAs) for IoT devices, allowing the devices to sign transactions themselves. The BCA of the IoT device is generated using encrypted PUF data, enabling other participants in the blockchain to verify the manufacturer, owner, and user of the device, as well as the actions and data they are taking.

Anomaly detection [185] is a crucial aspect of securing IoT networks, and machine learning techniques have been proposed to detect and classify anomalies in these networks. Various machine learning algorithms, including Logistic Regression [186], Support Vector Machine, Decision Tree [187], Random Forest, and Artificial Neural Networks, have been tested for detecting IoT network attacks, with Random Forest techniques performing relatively well. Federated Learning (FL) [188] is a distributed collaborative AI approach that enables data training without directly sharing user data with third parties. Using a central server to coordinate the training of multiple devices, FL enhances privacy while supporting the collaborative training of shared global models, which benefits network operators and IoT users by conserving network resources. When combined with blockchain technology, homomorphic encryption can enhance privacy, security, and trust in the processing and storage of sensitive data generated by IoT devices. However, as Shrestha and Kim [189] point out, these technologies alone are insufficient to provide comprehensive security for the IoT ecosystem. Further research is needed to address the challenges and opportunities posed by their integration.

In addition to technological measures, the design of IoT ecosystems can also enhance security and privacy. For example, in selecting and distributing sensors in smart architecture, high-dimensional sensors should be prioritized in public spaces, where there are many users with complex behaviors and low privacy levels, to obtain a wide range of intuitive and non-intrusive information with a small number of sensors. In private spaces, low-dimensional sensors can easily sense and understand user behavior through simple training, and the program logic can be relatively simple to manipulate. A large number of low-dimensional sensors can be distributed in the building space, so that each sensor can only obtain a limited range of data information types, which can protect users' personal and private information to a certain extent and enhance the personal information security of users.

Psychological safety issues are also important in IoT ecosystem design. Although the dynamic behavior of smart building interaction will not hurt users in theory, due to the lack of knowledge about the dynamic behavior of the building and the inherent distrust of machines, there may be a sense of lack of psychological security. To address this, smart devices should include reminder settings informing users of the data collection status. For example, a camera with a physical mask can help users know whether their data is being extracted, which is also conducive to helping users trust the Internet of Things.

Digital twins can also be utilized to increase user confidence in the security of their data. By creating a virtual replica of the physical system, users can monitor how their data is acquired, transferred, and used in real-time, enhancing their sense of control and trust in the system. For example, in a smart home system, using a digital twin can allow users to monitor the status of their security cameras, smart locks, and other devices in real time. By accessing the digital twin, users can view the data collected by the cameras and the lock logs to ensure that their home is secure and that only authorized individuals are entering the premises. In the context of a smart city, digital twin technology can also allow people to monitor how their personal information is being collected, transferred, and used for the purpose of city development. By providing users with transparency into the data ecosystem, including how their personal information is being used and by whom, digital twin technology can help enhance people's sense of security and trust in the system.

Moreover, by implementing an award system that recognizes and rewards people for their contributions to the smart city ecosystem, citizens can be incentivized greater engagement and participation. For example, by offering rewards such as discounts on city services or other benefits, people may be more likely to actively engage with the system, thereby contributing to the development and success of the smart city.

5. Conclusions

The emphasis on understanding the challenges associated with the Internet of Things (IoT) is critical for researchers, practitioners, and policymakers, as it helps identify the gaps in the development and implementation of IoT. As this study elucidates, a comprehensive grasp of these challenges can pave the way for more informed and targeted solutions. A deeper understanding of the technical, social, and economic challenges will not only facilitate the development of IoT technologies that can contribute to enhanced living environments but also enable the creation of effective strategies and policies to guide their adoption.

It is crucial to consider IoT development within an expansive framework, recognizing the potential of smart architecture and urban planning to act as platforms for IoT devices, as this perspective is integral to the future enhancement of our living environments. These platforms can serve as the foundation for integrating various IoT systems, fostering the creation of an intelligent ecosystem that can enhance living environments. This approach not only underscores the need for interoperability among diverse IoT devices but also emphasizes the role of smart urban planning and architectural design in the successful deployment of IoT technologies.

During this research, we have proposed the intelligent IoT ecosystem as a viable solution to mitigate some of these challenges. Through the utilization of advanced data analysis and machine learning algorithms, the ecosystem can effectively integrate multiple sources of data, enabling real-time monitoring and control of IoT devices. This holistic approach would lead to an improved user experience, offering personalized services, increasing efficiency, and ensuring cost savings, while enhancing security and privacy. The implications of these findings are profound for the future trajectory of IoT and smart cities research. Continued focus on technical and social challenges, including improving data privacy and security, compatibility, and interoperability, as well as addressing ethical and legal considerations, is paramount. Additionally, a reinforced commitment to investment in IoT and smart cities will help overcome economic challenges and realize the untapped benefits of these technologies.

In conclusion, while the potential of IoT to significantly enhance living environments is undeniable, it is equally important to address the multifaceted challenges that accompany its deployment. Sustained investment in research and development, an ongoing commitment to overcoming the identified challenges, and active engagement with stakeholders across various sectors, can ensure that the benefits of IoT are realized by all. Future research recommendations include a continued focus on these areas, ensuring that the advantages of IoT and smart cities are equitably distributed and that these technologies are developed and deployed in an inclusive, accessible, and ethical manner.

Acknowledgments

This research is funded and supported by the Shenzhen Science and Technology Innovation Committee (WDZC20200822215113001) and the Fundamental Research Funds for the Central Universities (No. 2023RC04).

References

[1] A. Zielonka, A. Sikora, M. Woźniak, W. Wei, Q. Ke, Z. Bai, Intelligent internet of things system for smart home optimal convection, IEEE Trans. Industr. Inform. 17 (6) (2020) 4308–4317.

[2] C. Stolojescu-Crisan, C. Crisan, B.-P. Butunoi, An iot-based smart home automation system, Sensors 21 (11) (2021) 3784.

[3] M. Mohammed, S. Desyansah, S. Al-Zubaidi, E. Yusuf, An internet of things-based smart homes and healthcare monitoring and management system, J. Phys. Conf. Ser. 1450 (2020) 012079. IOP Publishing.

[4] M. Usak, M. Kubiatko, M.S. Shabbir, O. Viktorovna Dudnik, K. Jermsittiparsert, L. Rajabion, Health care service delivery based on the internet of things: a systematic and comprehensive study, Int. J. Commun. Syst. 33 (2) (2020) e4179.

[5] S. Zeadally, A.K. Das, N. Sklavos, Cryptographic technologies and protocol standards for internet of things, Internet Things 14 (2021) 100075.

[6] A. Khanna, R. Goyal, M. Verma, D. Joshi, Intelligent traffic management system for smart cities, in: Futuristic Trends in Network and Communication Technologies: First International Conference, FTNCT 2018, Solan, India, February 9–10, 2018, Revised Selected Papers 1, Springer, 2019, pp. 152–164.

[7] T. Alam, M.A. Khan, N.K. Gharaibeh, M.K. Gharaibeh, Big data for smart cities: a case study of Neom City, Saudi Arabia, in: Smart Cities: A Data Analytics Perspective, Springer, Cham, 2021, pp. 215–230.

[8] C.K. Metallidou, K.E. Psannis, E.A. Egyptiadou, Energy efficiency in smart buildings: Iot approaches, IEEE Access 8 (2020) 63679–63699.

[9] A. Murthy, D. Han, D. Jiang, T. Oliveira, Lighting-enabled smart city applications and ecosystems based on the Iot, in: 2015 IEEE 2nd World Forum on Internet of Things (WF-IoT), IEEE, 2015, pp. 757–763.

[10] T. Ali, M. Irfan, A.S. Alwadie, A. Glowacz, Iot-based smart waste bin monitoring and municipal solid waste management system for smart cities, Arab. J. Sci. Eng. 45 (2020) 10185–10198.

[11] P. Saarika, K. Sandhya, T. Sudha, S.T.S.U. Iot, International conference on smart technologies for smart nation (SmartTechCon), IEEE 2017 (2017) 1104–1107.

[12] U. Osisiogu, A Review on Cyber-Physical Security of Smart Buildings and Infrastructure, in: 2019 15th International Conference on Electronics, Computer and Computation (ICECCO), IEEE, 2019, pp. 1–4.

[13] F. Al-Turjman, H. Zahmatkesh, R. Shahroze, An overview of security and privacy in smart cities' iot communications, Trans. Emerg. Telecommun. Technol. 33 (3) (2022) e3677.

[14] H.A. Al-Mohammed, E. Yaacoub, IEEE, On the use of quantum communications for securing iot devices in the 6g era, in: IEEE International Conference on Communications (ICC), Electr Network, 2021.

[15] C. Jenila, R.K. Jeyachitra, Green indoor optical wireless communication systems: pathway towards pervasive deployment, Digit. Commun. Netw. 7 (3) (2021) 410–444.

[16] X. Meng, X. Yu, W. Chen, Y. Zhao, J. Zhang, Residual-adaptive key provisioning in quantum-key-distribution enhanced Internet of Things (Q-IoT), 2020 International Wireless Communications and Mobile Computing (IWCMC), Limassol, Cyprus, 2020, pp. 2022–2027.

[17] R. Lu, A new communication-efficient privacy-preserving range query scheme in fog-enhanced Iot, IEEE Internet Things J. 6 (2) (2019) 2497–2505.

[18] M. Shen, B. Ma, L. Zhu, X. Du, K. Xu, Secure phrase search for intelligent processing of encrypted data in cloud-based Iot, IEEE Internet Things J. 6 (2) (2019) 1998–2008.

[19] C. Zhou, A. Fu, S. Yu, W. Yang, H. Wang, Y. Zhang, Privacy-preserving federated learning in fog computing, IEEE Internet Things J. 7 (11) (2020) 10782–10793.

[20] J. Kang, Z. Xiong, D. Niyato, D. Ye, D.I. Kim, J. Zhao, Toward secure blockchain-enabled internet of vehicles: optimizing consensus management using reputation and contract theory, IEEE Trans. Veh. Technol. 68 (3) (2019) 2906–2920.

[21] J. Kang, R. Yu, X. Huang, M. Wu, S. Maharjan, S. Xie, Y. Zhang, Blockchain for secure and efficient data sharing in vehicular edge computing and networks, IEEE Internet Things J. 6 (3) (2019) 4660–4670.

[22] U. Hayat, N.A. Azam, A novel image encryption scheme based on an elliptic curve, Signal Process. 155 (2019) 391–402.

[23] Y. Luo, X. Ouyang, J. Liu, L. Cao, An image encryption method based on elliptic curve elgamal encryption and chaotic systems, IEEE Access 7 (2019) 38507–38522.

[24] W. Jang, S.-Y. Lee, Partial image encryption using format-preserving encryption in image processing systems for internet of things environment, Int. J. Distrib. Sens. Netw. 16 (3) (2020) 1550147720914779.

[25] A. Perez-Resa, M. Garcia-Bosque, C. Sanchez-Azqueta, S. Celma, A new method for format preserving encryption in high-data rate communications, IEEE Access 8 (2020) 21003–21016.

[26] C.-M. Chen, Y. Huang, K.-H. Wang, S. Kumari, M.-E. Wu, A secure authenticated and key exchange scheme for fog computing, Enterprise Inform. Syst. 15 (9) (2021) 1200–1215.

[27] M. Wazid, A.K. Das, V. Odelu, N. Kumar, W. Susilo, Secure remote user authenticated key establishment protocol for smart home environment, IEEE Trans. Depend. Secure Comput. 17 (2) (2020) 391–406.

[28] M. Angel Murillo-Escobar, M. Omar Meranza-Castillon, R. Martha Lopez-Gutierrez, C. Cruz-Hernandez, Suggested integral analysis for chaos-based image cryptosystems, Entropy 21 (8) (2019).

[29] C.L. Chowdhary, P.V. Patel, K.J. Kathrotia, M. Attique, K. Perumal, M.F. Ijaz, Analytical study of hybrid techniques for image encryption and decryption, Sensors 20 (18) (2020).

[30] A. Dorri, S.S. Kanhere, R. Jurdak, P. Gauravaram, Lsb: a lightweight scalable blockchain for iot security and anonymity, J. Parallel Distrib. Comput. 134 (2019) 180–197.

[31] P. Gope, B. Sikdar, Lightweight and privacy-preserving two-factor authentication scheme for Iot devices, IEEE Internet Things J. 6 (1) (2019) 580–589.

[32] S. Tuli, R. Mahmud, S. Tuli, R. Buyya, Fogbus: a blockchain-based lightweight framework for edge and fog computing, J. Syst. Softw. 154 (2019) 22–36.

[33] T.M. Fernandez-Carames, From pre-quantum to post-quantum iot security: a survey on quantum-resistant cryptosystems for the internet of things, IEEE Internet Things J. 7 (7) (2020) 6457–6480.

[34] W. Jiang, H. Li, G. Xu, M. Wen, G. Dong, X. Lin, Ptas: privacy-preserving thin-client authentication scheme in blockchain-based Pki, Future Gener. Comput. Syst. Int. J. eSci. 96 (2019) 185–195.

[35] H. Xiong, Y. Zhao, Y. Hou, X. Huang, C. Jin, L. Wang, S. Kumari, Heterogeneous signcryption with equality test for iiot environment, IEEE Internet Things J. 8 (21) (2021) 16142–16152.

[36] S.K. Sood, Mobile fog based secure cloud-Iot framework for enterprise multimedia security, Multimed. Tools Appl. 79 (15–16) (2020) 10717–10732.

[37] H.H. Mahmoud, A.S. Alghawli, M.K.M. Al-shammari, G.A. Amran, K.H. Mutmbak, K.H. Al-harbi, M.A.A. Al-qaness, Iot-based motorbike ambulance: secure and efficient transportation, Electronics 11 (18) (2022).

[38] A. Liu, A. Alqazzaz, H. Ming, B. Dharmalingam, Iotverif: automatic verification of Ssl/Tls certificate for Iot applications, IEEE Access 9 (2021) 27038–27050.

[39] P. Li, J. Su, X. Wang, Itls: lightweight transport-layer security protocol for Iot with minimal latency and perfect forward secrecy, IEEE Internet Things J. 7 (8) (2020) 6828–6841.

[40] R. Holz, J. Hiller, J. Amann, A. Razaghpanah, T. Jost, N. Vallina-Rodriguez, O. Hohlfeld, Tracking the deployment of Tls 1.3 on the web: a story of experimentation and centralization, ACM Sigcomm Comput. Commun. Rev. 50 (3) (2020) 4–15.

[41] P.M. Kumar, U.D. Gandhi, Enhanced Dtls with coap-based authentication scheme for the internet of things in healthcare application, J. Supercomput. 76 (6) (2020) 3963–3983.

[42] U. Banerjee, A. Wright, C. Juvekar, M. Waller, A.P.C. Arvind, An energy-efficient reconfigurable dtls cryptographic engine for securing internet-of-things applications, IEEE J. Solid State Circuits 54 (8) (2019) 2339–2352.

[43] A. Lohachab, Karambir., Ecc based inter-device authentication and authorization scheme using Mqtt for Iot networks, J. Inf. Secur. Appl. 46 (2019) 1–12.

[44] D. Dinculeana, X. Cheng, Vulnerabilities and limitations of mqtt protocol used between Iot devices, Appl. Sci. Basel 9 (5) (2019).

[45] R.H. Randhawa, A. Hameed, A.N. Mian, Energy efficient cross-layer approach for object security of coap for Iot devices, Ad. Hoc. Netw. 92 (2019) 101761.

[46] R. Herrero, Dynamic coap mode control in real time wireless Iot networks, IEEE Internet Things J. 6 (1) (2019) 801–807.

[47] G. Wang, Y. Sun, J. Chen, Y. Jiao, Z. Chang, H. Yu, C. Lin, G. Zhao, Quinoa Traceable System Based on Internet of Things, in: 11th IFIP WG 5.14 International Conference on Computer and Computing Technologies in Agriculture (CCTA), Jilin, Peoples R China, 2017, pp. 1–8.

[48] Z. Lv, B. Hu, H. Lv, Infrastructure monitoring and operation for smart cities based on Iot system, IEEE Trans. Industr. Inform. 16 (3) (2020) 1957–1962.

[49] V. Bianchi, P. Ciampolini, I. De Munari, Rssi-based indoor localization and identification for Zigbee wireless sensor networks in smart homes, IEEE Trans. Instrum. Meas. 68 (2) (2019) 566–575.

[50] S. Sinche, D. Raposo, N. Armando, A. Rodrigues, F. Boavida, V. Pereira, J.S. Silva, A survey of iot management protocols and frameworks, IEEE Commun. Surv. Tutor. 22 (2) (2020) 1168–1190.

[51] A. Mavromatis, C. Colman-Meixner, A.P. Silva, X. Vasilakos, R. Nejabati, D. Simeonidou, A software-defined iot device management framework for edge and cloud computing, IEEE Internet Things J. 7 (3) (2020) 1718–1735.

[52] S.S. Basu, J. Haxhibeqiri, M. Baert, B. Moons, A. Karaagac, P. Crombez, P. Camerlynck, J. Hoebeke, An end-to-end Lwm2m-based communication architecture for multimodal Nb-Iot/Ble devices, Sensors 20 (8) (2020).

[53] G. Glissaa, A. Meddeb, 6lowpsec: an end-to-end security protocol for 6lowpan, Ad Hoc Netw. 82 (2019) 100–112.

[54] H.A.A. Al-Kashoash, H. Kharrufa, Y. Al-Nidawi, A.H. Kemp, Congestion control in wireless sensor and 6lowpan networks: toward the internet of things, Wirel. Netw 25 (8) (2019) 4493–4522.

[55] U. Nguyen Quoc, N. Vu Hoai, A comparison of Amqp and Mqtt protocols for internet of things, in: 6th National-Foundation-for Science-and-Technology-Development (NAFOSTED) Conference on Information and Computer Science (NICS), Posts & Telecommunicat Inst Technol Hanoi, Hanoi, Vietnam, 2019, pp. 292–297.

[56] D. Uroz, R.J. Rodriguez, Characterization and evaluation of iot protocols for data exfiltration, IEEE Internet Things J. 9 (19) (2022) 19062–19072.

[57] P.J. Taylor, T. Dargahi, A. Dehghantanha, R.M. Parizi, K.-K.R. Choo, A systematic literature review of blockchain cyber security, Digit. Commun. Netw. 6 (2) (2020) 147–156.

[58] K. Yu, L. Tan, M. Aloqaily, H. Yang, Y. Jararweh, Blockchain-enhanced data sharing with traceable and direct revocation in Iiot, IEEE Trans. Industr. Inform. 17 (11) (2021) 7669–7678.

[59] M. Andoni, V. Robu, D. Flynn, S. Abram, D. Geach, D. Jenkins, P. McCallum, A. Peacock, Blockchain technology in the energy sector: a systematic review of challenges and opportunities, Renew. Sustain. Energy Rev. 100 (2019) 143–174.

[60] B. Chatterjee, D. Das, S. Maity, S. Sen, Rf-Puf: enhancing Iot security through authentication of wireless nodes using in-situ machine learning, IEEE Internet Things J. 6 (1) (2019) 388–398.

[61] R.W. Ahmad, K. Salah, R. Jayaraman, I. Yaqoob, M. Omar, Blockchain for waste management in smart cities: a survey, IEEE Access 9 (2021) 131520–131541.

[62] A.U.R. Khan, R.W. Ahmad, A blockchain-based Iot-enabled E-waste tracking and tracing system for smart cities, IEEE Access 10 (2022) 86256–86269.

[63] B. Bera, S. Saha, A.K. Das, A.V. Vasilakos, Designing blockchain-based access control protocol in Iot-enabled smart-grid system, IEEE Internet Things J. 8 (7) (2021) 5744–5761.

[64] A. Hasankhani, S.M. Hakimi, M. Bisheh-Niasar, M. Shafie-khah, H. Asadolahi, Blockchain technology in the future smart grids: a comprehensive review and frameworks, Int. J. Electr. Power Energy Syst. 129 (2021).

[65] A. Shahzad, K. Zhang, A. Gherbi, Privacy-preserving smart grid traceability using blockchain over iot connectivity, in: Proceedings of the 36th Annual ACM Symposium on Applied Computing, Association for Computing Machinery, 2021, pp. 699–706. https://doi.org/10.1145/3412841.3441949.

[66] J. Pan, J. Wang, A. Hester, I. Algerm, Y. Liu, Y. Zhao, Edgechain: an edge-Iot framework and prototype based on blockchain and smart contracts, IEEE Internet Things J. 6 (3) (2019) 4719–4732.

[67] Q. Kong, R. Lu, F. Yin, S. Cui, Blockchain-based privacy-preserving driver monitoring for maas in the vehicular Iot, IEEE Trans. Veh. Technol. 70 (4) (2021) 3788–3799.

[68] S.R. Niya, E. Schiller, I. Cepilov, F. Maddaloni, K. Aydinli, T. Surbeck, T. Bocek, B. Stiller, IEEE, Adaptation of proof-of-stake-based blockchains for Iot data streams, in: 1st IEEE International Conference on Blockchain and Cryptocurrency (IEEE ICBC), Seoul, South Korea, 2019, pp. 15–16.

[69] X. Fan, Q. Chai, M. Assoc Comp, Roll-Dpos (Sic): a randomized delegated proof of stake scheme for scalable blockchain-based internet of things systems, in: 15th EAI International Conference on Mobile and Ubiquitous Systems—Computing, Networking and Services (Mobiquitous), New York City, NY, 2018, pp. 482–484.

[70] Y. Wei, L. Liang, B. Zhou, X. Feng, IEEE, A modified blockchain dpos consensus algorithm based on anomaly detection and reward-punishment, in: 13th IEEE International Conference on Communication Software and Networks (ICCSN), Chongqing, Peoples R China, 2021, pp. 283–288.

[71] R. Cole, L. Cheng, IEEE Modeling the energy consumption of blockchain consensus algorithms, in: IEEE International Congress on Cybermatics/IEEE Conferences on Internet of Things, Green Computing and Communications, Cyber, Physical and Social Computing, Smart Data, Blockchain, Computer and Information Technology, Halifax, Canada, 2018, pp. 1691–1696.

[72] P.K. Sharma, N. Kumar, J.H. Park, Blockchain technology toward green Iot: opportunities and challenges, IEEE Netw. 34 (4) (2020) 263–269.

[73] A. Hakiri, B. Sellami, S. Ben Yahia, P. Berthou, A blockchain architecture for Sdn-enabled tamper-resistant Iot networks, 2020 Global Information Infrastructure and Networking Symposium (GIIS), Tunis, Tunisia, 2020, pp. 1–4. https://doi.org/10.1109/GIIS50753.2020.9248492.

[74] S. Aggarwal, N. Kumar, Cryptographic consensus mechanisms, in: S. Aggarwal, N. Kumar, P. Raj (Eds.), Blockchain Technology for Secure and Smart Applications across Industry Verticals, 2021, pp. 211–226.

[75] J. Suhan, W. Jie, A game-theoretic approach to storage offloading in Poc-based mobile blockchain mining, Proceedings of the Twenty-First International Symposium on Theory, Algorithmic Foundations, and Protocol Design for Mobile Networks and Mobile Computing, Association for Computing Machinery, 2020, pp. 171–180. https://doi.org/10.1145/3397166.3409136.

[76] L. Huang, S. Bi, Y.-J.A. Zhang, Deep reinforcement learning for online computation offloading in wireless powered mobile-edge computing networks, IEEE Trans. Mob. Comput. 19 (11) (2020) 2581–2593.

[77] R. Kitchin, M. Dodge, The (in)security of smart cities: vulnerabilities, risks, mitigation, and prevention, J. Urban Technol. 26 (2) (2019) 47–65.

[78] M. Muzammal, Q. Qu, B. Nasrulin, Renovating blockchain with distributed databases: an open source system, Future Gener. Comput. Syst. Int. J. eSci. 90 (2019) 105–117.

[79] Y. Wu, H. Huang, N. Wu, Y. Wang, M.Z.A. Bhuiyan, T. Wang, An incentive-based protection and recovery strategy for secure big data in social networks, Inf. Sci. 508 (2020) 79–91.

[80] G. He, G. Yang, Y. Cai, C. Ma, Design and implementation of traffic incident acquisition and reporting device based on Lte communication, J. Phys. Conf. Ser. 1486 (2020). 022023 (022025 pp.)-022023 (022025 pp.).

[81] D. Rodriguez-Garcia, V. Garcia-Diaz, C. Gonzalez Garcia, Crowdsl: platform for incidents management in a smart city context, Big Data Cogn. Comput. 5 (3) (2021).

[82] M.N. Aman, U. Javaid, B. Sikdar, Security function virtualization for Iot applications in 6g networks, IEEE Commun. Stand. Mag. 5 (3) (2021) 90–95.

[83] W.G. Hatcher, C. Qian, F. Liang, W. Liao, E.P. Blasch, W. Yu, Secure Iot search engine: survey, challenges issues, case study, and future research direction, IEEE Internet Things J. 9 (18) (2022) 16807–16823.

[84] H. Wu, D. Sun, L. Peng, Y. Yao, J. Wu, Q.Z. Sheng, Y. Yan, Dynamic edge access system in Iot environment, IEEE Internet Things J. 7 (4) (2020) 2509–2520.

[85] T. Kim, H. Kim, A design of automated vulnerability information management system for secure use of internet-connected devices based on internet-wide scanning methods, IEICE Trans. Inf. Syst. E104D (11) (2021) 1805–1813.

[86] T. Zitta, M. Lucki, L. Vojtech, M. Neruda, L. Mejzrova, Experimental load test statistics for the selected ips tools on low-performance Iot devices, J. Electr. Eng. 70 (4) (2019) 285–294.

[87] R. Basir, S. Qaisar, M. Ali, M. Naeem, Cloudlet selection in cache-enabled fog networks for latency sensitive Iot applications, IEEE Access 9 (2021) 93224–93236.

[88] Q. Liu, W. Ijntema, A. Drif, P. Pawelczak, M. Zuniga, K.S. Yildirim, Perpetual bluetooth communications for the Iot, IEEE Sensors J. 21 (1) (2021) 829–837.

[89] X. Yu, K. Ergun, L. Cherkasova, T.S. Rosing, Optimizing sensor deployment and maintenance costs for large-scale environmental monitoring, IEEE Trans. Comput. Aid. Des. Integr. Circuits Syst. 39 (11) (2020) 3918–3930.

[90] S.A. Mostafa, S.S. Gunasekaran, A. Mustapha, M.A. Mohammed, W.M. Abduallah, Modelling an adjustable autonomous multi-agent internet of things system for elderly smart home, in: 10th International Conference on Applied Human Factors and Ergonomics (AHFE)/AHFE International Conference on Neuroergonomics and Cognitive Engineering/AHFE International Conference on Industrial Cognitive Ergonomics and Engineering Psychology, Washington, DC, 2019, pp. 301–311.

[91] H. Moudoud, S. Cherkaoui, L. Khoukhi, IEEE, Towards a scalable and trustworthy blockchain: Iot use case, in: IEEE International Conference on Communications (ICC), Electr Network, 2021.

[92] Y. Li, X. Huang, S. Wang, Multiple protocols interworking with open connectivity foundation in in fog networks, IEEE Access 7 (2019) 60764–60773.

[93] H.S. Oh, W.S. Jeon, D.G. Jeong, Ocf bridging techniques for Uwb/Lora Iot ecosystems, IEEE Access 10 (2022) 58845–58857.

[94] O. Tomanek, L. Kenel, IEEE, Security and privacy of using alljoyn iot framework at home and beyond, in: 2nd International Conference on Intelligent Green Building and Smart Grid (IGBSG), Prague, Czech Republic, 2016, pp. 18–23.

[95] M. Noura, M. Atiquzzaman, M. Gaedke, Interoperability in Internet of things: taxonomies and open challenges, Mobile Netw. Appl. 24 (3) (2019) 796–809.

[96] A.V. Brega, O.V. Erokhina, Smart city transport technologies: infrastructure of the future, 2022 Systems of Signals Generating and Processing in the Field of on Board Communications, Russian Federation, Moscow, 2022, pp. 1–5. https://doi.org/10.1109/IEEECONF53456.2022.9744352.

[97] M. Menendez, L. Ambuhl, Implementing design and operational measures for sustainable mobility: lessons from Zurich, Sustainability 14 (2) (2022).

[98] X. Wang, S. Shen, D. Bezzina, J.R. Sayer, H.X. Liu, Y. Feng, Data infrastructure for connected vehicle applications, Transp. Res. Rec. 2674 (5) (2020) 85–96.

[99] A. Brutti, P. De Sabbata, A. Frascella, N. Gessa, R. Ianniello, C. Novelli, S. Pizzuti, G. Ponti, Smart city platform specification: a modular approach to achieve interoperability in smart cities, in: F. Cicirelli, A. Guerrieri, C. Mastroianni, G. Spezzano, A. Vinci (Eds.), Internet of Things for Smart Urban Ecosystems, 2019, pp. 25–50.

[100] I. Tsampoulatidis, N. Komninos, E. Syrmos, D. Bechtsis, Universality and interoperability across smart city ecosystems, in: 10th International Conference on Distributed, Ambient and Pervasive Interactions (DAPI) Held as Part of the 24th International Conference on Human-Computer Interaction (HCII), Electr Network, 2022, pp. 218–230.

[101] V. Araujo, K. Mitra, S. Saguna, C. Ahlund, Performance evaluation of fiware: a cloud-based iot platform for smart cities, J. Parallel Distrib. Comput. 132 (2019) 250–261.

[102] M. Wazid, A.K. Das, V.K. Bhat, A.V. Vasilakos, Lam-Ciot: lightweight authentication mechanism in cloud-based Iot environment, J. Netw. Comput. Appl. 150 (2020).

[103] V.L. Camacho, E.d.l. Guia, T. Olivares, M.J. Flores, L. Orozco-Barbosa, Data capture and multimodal learning analytics focused on engagement with a new wearable Iot approach, IEEE Trans. Learn. Technol. 13 (4) (2020) 704–717.

[104] T.M. Fernandez-Carames, P. Fraga-Lamas, Towards next generation teaching, learning, and context-aware applications for higher education: a review on blockchain, Iot, fog and edge computing enabled smart campuses and universities, Appl. Sci. 9 (21) (2019) 4479. https://doi.org/10.3390/app9214479.

[105] M.A. Ferrag, M. Derdour, M. Mukherjee, A. Derhab, L. Maglaras, H. Janicke, Blockchain technologies for the internet of things: research issues and challenges, IEEE Internet Things J. 6 (2) (2019) 2188–2204.

[106] H. Felzmann, E.F. Villaronga, C. Lutz, A. Tamò-Larrieux, Transparency you can trust: transparency requirements for artificial intelligence between legal norms and contextual concerns, Big Data Soc. 6 (1) (2019). 2053951719860542.

[107] L. Malina, G. Srivastava, P. Dzurenda, J. Hajny, S. Ricci, A privacy-enhancing frame-work for internet of things services, network and system security, in: 13th International Conference, NSS 2019, Sapporo, Japan, December 15–18, 2019, Proceedings 13, Springer, 2019, pp. 77–97.

[108] WHO, the UNICEF, Global Report on Assistive Technology, WHO and the UNICEF, 2022.

[109] L. Coetzee, G. Olivrin, Inclusion through the internet of things, Assist. Technol. (2012) 953–978.

[110] S. Abou-Zahra, J. Brewer, M. Cooper, Web standards to enable an accessible and inclusive internet of things (Iot), in: Proceedings of the 14th International Web for All Conference, 2017, pp. 1–4.

[111] J. Rochford, Accessibility and IoT/smart and connected communities, AIS Trans. Hum.-Comput. Interact. 11 (4) (2019) 253–263.

[112] P. Blanck, Equality: the right to the web, in: Routledge Handbook of Disability Law and Human Rights, Routledge, 2016, pp. 166–194.

[113] L. Qiao, A. Sullivan, Twine screen reader: a browser extension for improving the accessibility of twine stories for people with visual impairments, in: 15th International Conference on Interactive Digital Storytelling (ICIDS), Univ California, Santa Cruz, CA, 2022, pp. 577–589.

[114] A. Schnack, M.J. Wright, J.L. Holdershaw, Immersive virtual reality technology in a three-dimensional virtual simulated store: investigating telepresence and usability, Food Res. Int. 117 (2019) 40–49.

[115] G. Li, L. Zhang, Y. Sun, J. Kong, Towards the Semg Hand: internet of things sensors and haptic feedback application, Multimed. Tools Appl. 78 (21) (2019) 29765–29782.

[116] D. Sayassatov, N. Cho, The influence of learning styles on a model of Iot-based inclusive education and its architecture, J. Inform. Technol. Appl. Manag. 26 (5) (2019) 27–39.

[117] J. Rodrigues, A. Cardoso, Blockchain in smart cities: an inclusive tool for persons with disabilities, in: 2019 Smart City Symposium Prague (SCSP), IEEE, 2019, pp. 1–6.

[118] K.-A. Shim, Design principles of secure certificateless signature and aggregate signature schemes for Iot environments, IEEE Access 10 (2022) 124848–124857.

[119] S. Wang, K. Bolling, W. Mao, J. Reichstadt, D. Jeste, H.-C. Kim, C. Nebeker, Technology to support aging in place: older adults' perspectives, Healthcare 7 (2) (2019).

[120] M.A. Kuhail, S. Farooq, R. Hammad, M. Bahja, Characterizing visual programming approaches for end-user developers: a systematic review, IEEE Access 9 (2021) 14181–14202.

[121] J.S. de Oliveira Neto, Inclusive Smart Cities: Theory and Tools to Improve the Experience of People with Disabilities in Urban Spaces, Université Paris Saclay (COmUE); Universidade de São Paulo (Brésil), 2018.

[122] A. Brunete, E. Gambao, M. Hernando, R. Cedazo, Smart assistive architecture for the integration of Iot devices, robotic systems, and multimodal interfaces in healthcare environments, Sensors 21 (6) (2021) 2212.

[123] L.a. Tawalbeh, F. Muheidat, M. Tawalbeh, M. Quwaider, Iot privacy and security: challenges and solutions, Appl. Sci. 10 (12) (2020) 4102.

[124] P. Menard, G.J. Bott, Analyzing Iot users' mobile device privacy concerns: extracting privacy permissions using a disclosure experiment, Comput. Secur. 95 (2020) 101856.

[125] M.A. Al-Garadi, A. Mohamed, A.K. Al-Ali, X. Du, I. Ali, M. Guizani, A survey of machine and deep learning methods for internet of things (Iot) security, IEEE Commun. Surv. Tutor. 22 (3) (2020) 1646–1685.

[126] C. Boje, A. Guerriero, S. Kubicki, Y. Rezgui, Towards a semantic construction digital twin: directions for future research, Autom. Constr. 114 (2020).

[127] I. Mistry, S. Tanwar, S. Tyagi, N. Kumar, Blockchain for 5g-enabled iot for industrial automation: a systematic review, solutions, and challenges, Mech. Syst. Signal Process. 135 (2020).

[128] C.J. Hoofnagle, B. van der Sloot, F.Z. Borgesius, The European Union general data protection regulation: what it is and what it means, Inform. Commun. Technol. Law 28 (1) (2019) 65–98.

[129] R.S. Burnside, The electronic communications privacy act of 1986: the challenge of applying ambiguous statutory language to intricate telecommunication technologies, Rutgers Comput. Technol. Law J. 13 (2) (1987) 451–517.

[130] S. Taylor, Protecting privacy in Canada's private sector, Inf. Manag. J. 37 (4) (2003) 33–39.

[131] M. Vermanen, M.M. Rantanen, V. Harkke, Ethical framework for Iot deployment in smes: individual perspective, Internet Res. 32 (7) (2021) 185–201.

[132] G. Adamson, J.C. Havens, R. Chatila, Designing a value-driven future for ethical autonomous and intelligent systems, Proc. IEEE 107 (3) (2019) 518–525.

[133] P. Radanliev, D.C. De Roure, R. Nicolescu, M. Huth, R.M. Montalvo, S. Cannady, P. Burnap, Future developments in cyber risk assessment for the internet of things, Comput. Ind. 102 (2018) 14–22.

[134] S. Yasar, Sustainable financing of smart cities, Artif. Intel. Perspect. Smart Cities (2022) 155–184.

[135] T. Völker, Z. Kovacic, R. Strand, Indicator development as a site of collective imagination? The case of european commission policies on the circular economy, Cult. Organ. 26 (2) (2020) 103–120.

[136] L.A. Colombo, M. Pansera, R. Owen, The discourse of eco-innovation in the European Union: an analysis of the eco-innovation action plan and horizon 2020, J. Clean. Prod. 214 (2019) 653–665.

[137] J. Engelbert, L. van Zoonen, F. Hirzalla, Excluding citizens from the european smart city: the discourse practices of pursuing and granting smartness, Technol. Forecast. Soc. Chang. 142 (2019) 347–353.

[138] E.A. Metwally, A.A. Farid, M.R. Ismail, Development of an Iot assessment method: an interdisciplinary framework for energy efficient buildings, Energ. Buildings 254 (2022) 111545.

[139] M.K. Nematchoua, S. Asadi, S. Reiter, Estimation, analysis and comparison of carbon emissions and construction cost of the two tallest buildings located in United States and China, Int. J. Environ. Sci. Technol. 19 (10) (2022) 9313–9328.

[140] E. Ismagilova, L. Hughes, Y.K. Dwivedi, K.R. Raman, Smart cities: advances in research-an information systems perspective, Int. J. Inf. Manag. 47 (2019) 88–100.

[141] T. Kalil, Public policy and the national information infrastructure, Bus. Econ. 30 (4) (1995) 15–20.

[142] J. Shah, J. Kothari, N. Doshi, a survey of smart city infrastructure via case study on new york, in: 10th International Conference on Emerging Ubiquitous Systems and Pervasive Networks (EUSPN)/9th International Conference on Current and Future Trends of Information and Communication Technologies in Healthcare (ICTH), Coimbra, Portugal, 2019, pp. 702–705.

[143] T. Yigitcanlar, H. Han, M. Kamruzzaman, G. Ioppolo, J. Sabatini-Marques, The making of smart cities: are Songdo, Masdar, Amsterdam, San Francisco and Brisbane the best we could build? Land Use Policy 88 (2019) 104187.

[144] P. Cardullo, R. Kitchin, Smart urbanism and smart citizenship: the neoliberal logic of 'citizen-focused' smart cities in Europe, Environ. Plan. C Polit. Space 37 (5) (2019) 813–830.

[145] Z.D.W. Putra, W.G. van der Knaap, Urban innovation system and the role of an open web-based platform: the case of amsterdam smart city, J. Reg. City Plan. 29 (3) (2018) 234–249.

[146] V. Roblek, 5—the Smart City of Vienna, in: L. Anthopoulos (Ed.), Smart City Emergence, Elsevier, 2019, pp. 105–127.

[147] T. Bakıcı, E. Almirall, J. Wareham, A smart city initiative: the case of Barcelona, J. Knowl. Econ. 4 (2013) 135–148.

[148] M. Yli-Ojanpera, S. Sierla, N. Papakonstantinou, V. Vyatkin, Adapting an agile manufacturing concept to the reference architecture model industry 4.0: a survey and case study, J. Industr. Inform. 15 (2019) 147–160.

[149] M.A. Ferrag, L. Maglaras, S. Moschoyiannis, H. Janicke, Deep learning for cyber security intrusion detection: approaches, datasets, and comparative study, J. Inform. Secur. Appl. (2020) 50.

[150] G. Manogaran, P.M. Shakeel, H. Fouad, Y. Nam, S. Baskar, N. Chilamkurti, R. Sundarasekar, Wearable Iot smart-log patch: an edge computing-based bayesian deep learning network system for multi access physical monitoring system, Sensors 19 (13) (2019).

[151] Y. Liu, S. Xie, Y. Zhang, Cooperative offloading and resource management for Uav-enabled mobile edge computing in power Iot system, IEEE Trans. Veh. Technol. 69 (10) (2020) 12229–12239.

[152] G. Zhu, D. Liu, Y. Du, C. You, J. Zhang, K. Huang, Toward an intelligent edge: wireless communication meets machine learning, IEEE Commun. Mag. 58 (1) (2020) 19–25.

[153] H. Li, X. Ding, Y. Yang, X. Huang, G. Zhang, IEEE, Spectrum occupancy prediction for internet of things via long short-term memory, in: IEEE International Conference on Consumer Electronics-Taiwan (IEEE ICCE-TW), Ilan, Taiwan, 2019.

[154] N.S. Akash, S. Rouf, S. Jahan, A. Chowdhury, A. Chakrabarty, J. Uddin, Botnet detection in iot devices using random forest classifier with independent component analysis, J. Inform. Commun. Technol. Malaysia 21 (2) (2022) 201–232.

[155] I. Zakariyya, M.O. Al-Kadri, H. Kalutarage, IEEE, Resource efficient boosting method for Iot security monitoring, in: IEEE 18th Annual Consumer Communications and Networking Conference (CCNC), Electr Network, 2021.

[156] A. Pandey, R. Vamsi, S. Kumar, Handling device heterogeneity and orientation using multistage regression for gmm based localization in Iot networks, IEEE Access 7 (2019) 144354–144365.

[157] W. Yang, J. Zhang, C. Wang, X. Mo, Situation prediction of large-scale internet of things network security, EURASIP J. Inf. Secur. 2019 (1) (2019).

[158] Z. Chen, Y. Leng, Y. Hu, An improved incomplete Ap clustering algorithm based on K nearest neighbours, Int. J. Embed. Syst. 11 (3) (2019) 269–277.

[159] M.A. Fouad, A.T. Abdel-Hamid, Ieee, On detecting iot power signature anomalies using hidden Markov model (Hmm), in: 31st IEEE International Conference on Microelectronics (ICM), Cairo, Egypt, 2019, pp. 108–112.

[160] D. Gupta, S. Rani, S.H. Ahmed, S. Verma, M.F. Ijaz, J. Shafi, Edge caching based on collaborative filtering for heterogeneous Icn-Iot applications, Sensors 21 (16) (2021).

[161] W. Li, X. Zhou, S. Shimizu, M. Xin, J. Jiang, H. Gao, Q. Jin, Personalization recommendation algorithm based on trust correlation degree and matrix factorization, IEEE Access 7 (2019) 45451–45459.

[162] Z. Zhou, X. Chen, E. Li, L. Zeng, K. Luo, J. Zhang, Edge intelligence: paving the last mile of artificial intelligence with edge computing, Proc. IEEE 107 (8) (2019) 1738–1762.

[163] L. Li, M. Guo, L. Ma, H. Mao, Q. Guan, Online workload allocation via fog-fog-cloud cooperation to reduce IoT task service delay, Sensors 19 (18) (2019) 3830. https://doi.org/10.3390/s19183830.

[164] A. Yousefpour, C. Fung, N. Tam, K. Kadiyala, F. Jalali, A. Niakanlahiji, J. Kong, J.P. Jue, All one needs to know about fog computing and related edge computing paradigms: a complete survey, J. Syst. Archit. 98 (2019) 289–330.

[165] N. Moustafa, A new distributed architecture for evaluating Ai-based security systems at the edge: network Ton_Iot datasets, Sustain. Cities Soc. 72 (2021).

[166] S. Xinjian, W. Lijie, Q. Xiaoyang, Y. Runhua, W. Yangyang, W. Dequan, L. Boxian, Deep reinforcement learning cloud-edge-terminal computation resource allocation mechanism for IoT, In: Q. Liu, X. Liu, T. Shen, X. Qiu (Eds.), The 10th International Conference on Computer Engineering and Networks. CENet 2020. Advances in Intelligent Systems and Computing, vol. 1274, Springer, Singapore. https://doi.org/10.1007/978-981-15-8462-6_177.

[167] M.S. Jassas, Q.H. Mahmoud, IEEE, Evaluation of failure analysis of iot applications using edge-cloud architecture, in: 16th Annual IEEE International Systems Conference (SysCon), Electr Network, 2022.

[168] B. Liu, C. Han, Research on wireless network virtualization positioning technology based on next-generation agile Iot technology, J. Interconnect. Netw. (2022).

[169] X. Guo, B. Tang, Security threats and countermeasures for software-defined Internet of Things, in: X. Sun, X. Zhang, Z. Xia, E. Bertino (Eds.), Advances in Artificial Intelligence and Security. ICAIS 2022. Communications in Computer and Information Science, vol. 1588, Springer, Cham, 2022. https://doi.org/10.1007/978-3-031-06764-8_51.

[170] I. Pittaras, G.C. Polyzos, I.C. Soc, (Poster) Smarttwins: secure and auditable Dlt-based digital twins for the Wot, in: 18th Annual International Conference on Distributed Computing in Sensor Systems (DCOSS), Los Angeles, CA, 2022, pp. 82–84.

[171] Z. Callejas, R. López-Cózar, Designing smart home interfaces for the elderly, ACM Sigaccess Access. Comput. 95 (2009) 10–16.

[172] P. Hess, S. Bracke, Smart material actuators as a contributor for Iot-based smart applications and systems: analyzing prototype and process measurement data of shape memory actuators for reliability and risk prognosis, J. Adv. Mech. Des. Syst. Manuf. 14 (2) (2020).

[173] C. Covaci, A. Gontean, Energy harvesting with piezoelectric materials for Iot—review, ITM Web Conf. 29 (2019). 03010 (03014 pp.)–03010 (03014 pp.).

[174] T. Zhuang, M. Ren, X. Gao, M. Dong, W. Huang, C. Zhang, Insulation condition monitoring in distribution power grid via Iot-based sensing network, IEEE Trans. Power Deliv. 34 (4) (2019) 1706–1714.

[175] A. Munoz-Arcentales, S. Lopez-Pernas, J. Conde, A. Alonso, J. Salvachua, J.J. Hierro, Enabling context-aware data analytics in smart environments: an open source reference implementation, Sensors 21 (21) (2021).

[176] X. Chen, Y. Liu, J. Zhang, Traffic prediction for internet of things through support vector regression model, Internet Technol. Lett. 5 (3) (2022).

[177] J. Yoon, Ieee, Using a Deep-Learning Approach for Smart Iot Network Packet Analysis, in: 4th IEEE European Symposium on Security and Privacy (EUROS&P), Royal Inst Technol, Stockholm, Sweden, 2019, pp. 291–299.

[178] R. Raman, A. Chirputkar, Ensemble learning method for improving the healthcare Iot system, Cardiometry 25 (2022) 171–177.

[179] S. Jothiraj, S. Balu, A Novel linear svm-based compressive collaborative spectrum sensing (Ccss) scheme for Iot cognitive 5g network, Soft. Comput. 23 (18) (2019) 8515–8523.

[180] S. Rawashdeh, W. Eyadat, A. Magableh, W. Mardini, M.B. Yasin, IEEE, Sustainable smart world, in: 10th International Conference on Information and Communication Systems (ICICS), Jordan Univ Sci & Technol, Irbid, Jordan, 2019, pp. 217–223.

[181] S. Zeadally, F.K. Shaikh, A. Talpur, Q.Z. Sheng, Design architectures for energy harvesting in the internet of things, Renew. Sustain. Energy Rev. 128 (2020).

[182] A. Babaei, G. Schiele, Physical unclonable functions in the internet of things: state of the art and open challenges, Sensors 19 (14) (2019).

[183] A. Shamsoshoara, A. Korenda, F. Afghah, S. Zeadally, A survey on physical unclonable function (Puf)-based security solutions for internet of things, Comput. Netw. 183 (2020).

[184] J. Arcenegui, R. Arjona, I. Baturone, Secure management of Iot devices based on blockchain non-fungible tokens and physical unclonable functions, in: Applied Cryptography and Network Security Workshops: ACNS 2020 Satellite Workshops, AIBlock, AIHWS, AIoTS, Cloud S&P, SCI, SecMT, and SiMLA, Rome, Italy, October 19–22, 2020, Proceedings 18, Springer, 2020, pp. 24–40.

[185] A.A. Cook, G. Misirli, Z. Fan, Anomaly detection for Iot time-series data: a survey, IEEE Internet Things J. 7 (7) (2020) 6481–6494.

[186] S. Das, The early bird catches the worm-first mover advantage through Iot adoption for Indian Public Sector Retail Oil outlets, J. Glob. Inf. Technol. Manag. 22 (4) (2019) 280–308.

[187] M.B. Taha, H. Suwi, F. Khaswneh, K. Alzaareer, Adaptive ciphertext policy attribute based encryption scheme for internet of things devices using decision tree, Revue d'Intelligence Artificielle 34 (3) (2020) 233–241.

[188] Q. Wu, K. He, X. Chen, Personalized federated learning for intelligent Iot applications: a cloud-edge based framework, IEEE Open J. Comput. Soc. 1 (2020) 35–44.

[189] R. Shrestha, S. Kim, Integration of Iot with blockchain and homomorphic encryption: challenging issues and opportunities, in: S. Kim, G.C. Deka, P. Zhang (Eds.), Role of Blockchain Technology in Iot Applications, 2019, pp. 293–331.

About the authors

Dr. Jie Liu is currently a lecturer in the School of Digital Media and Design at Beijing University of Posts and Telecommunications (BUPT). Her main research interests include Human–Architecture Interaction Design, Smart Architecture, and spatial design for Intelligent Buildings. Dr. Liu earned her PhD in Architecture and conducted postdoctoral research in the field of human-computer interaction at Tsinghua University. She supervised a Chinese Postdoctoral Science Foundation project and participated in several national grants and experimental projects. She has published 30 papers and received the Young CAADRIA Award in 2019. Additionally, she has contributed to numerous installation designs, architectural projects, and urban designs. Her works have been exhibited in international events such as the Milan International Design Triennale and Beijing International Design Week.

Hanyang Hu received a bachelor's degree in Architecture from Wuhan University, Wuhan, China in 2019 and a master's degree in Architecture from UC Berkeley, Berkeley, United States, in 2020. She is currently working towards a PhD degree in Architecture with the School of Architecture, Tsinghua University, Beijing, China. Her research interests include 3D concrete printing, human-robot interaction, and affective computing.

Dr. Dan Luo is an architect with strong computer science background. She has a PhD in digital design and fabrication from Tsinghua University (2019), a Master of Architecture from Columbia University (2014) and a Master of Computer and Information Technology from University of Pennsylvania (2023). Dr. Luo's expertise is in digital design and applied robotic construction. She has worked for pioneer design firm UNStudio. Her research combined advance fabrication technologies and design-build practice to explore the future of automated construction. Dr. Luo's work has showcased an unique combination of robotic automation and Artificial Intelligence to create end-end material computation methods that directly bridge between fabrication control and material performance. Not only yield in publications on top journals and conferences, Dr. Luo has also developed robotic system applied during the on-site construction of built projects.

Printed in the United States
by Baker & Taylor Publisher Services